Archives of the Scientific Revolution

The Formation and Exchange of
Ideas in Seventeenth-Century Europe

Muniment chest presented to the Royal Society by William Ball, Treasurer 1660–63. On resigning as Treasurer, Ball 'presented the Society with an iron chest having three locks and keys'. It has been in the Society's possession ever since.

Archives of the Scientific Revolution

The Formation and Exchange of Ideas in Seventeenth-Century Europe

EDITED BY

Michael Hunter

THE BOYDELL PRESS

First published 1998
The Boydell Press, Woodbridge

ISBN 0 85115 553 7

The Boydell Press is an imprint of Boydell & Brewer Ltd
PO Box 9, Woodbridge, Suffolk IP12 3DF, UK
and of Boydell & Brewer Inc.
PO Box 41026, Rochester, NY 14604–4126, USA

A catalogue record for this book is available
from the British Library

Library of Congress Cataloging-in-Publication data
Archives of the scientific revolution : the formation and exchange of ideas in seventeenth-century Europe / edited by Michael Hunter.
p. cm.
Includes bibliographical references and index.
ISBN 0–85115–553–7 (alk. paper)
1. Scientific archives – History. 2. Science – History – Sources.
I. Hunter, Michael Cyril William.
Q224.A73 1998
026.5–dc21 97–39325

This publication is printed on acid-free paper

Printed in Great Britain by
St Edmundsbury Press Ltd, Bury St Edmunds, Suffolk

Contents

Illustrations

Acknowledgments

This volume stems from a conference held at the Royal Society on 11–12 April 1996 entitled 'Archives of the Scientific Revolution'. On that occasion, oral versions were delivered of all but one of the papers included here, together with two other papers – one, an illustrated presentation on the Royal Society's archive by its archivist, Mary Sampson, and the other a paper on Elias Ashmole by David Harley, the inclusion of a revised version of which has regrettably been prevented by various unfortunate circumstances. James O'Hara's paper on the Leibniz Archiv did not form part of the conference, but Dr O'Hara was present, and he spoke informally about the Leibniz Archiv as part of a further component of the conference, a round-table discussion on editing scientific texts. The other participants in this were Michael Hunter, Mark Greengrass, Robert Goulding (who talked about his work on Sir Henry Savile), John Milton (speaking about John Locke), and Fritz Nagel from the Bernouilli-Edition, Basel. The discussion which took place was highly profitable, and it has informed the treatment of related issues in the Introduction that follows: it has not, however, seemed appropriate otherwise to record its outcome here. The conference was introduced by Professor John Rowlinson, and concluding remarks were given by Sir Alan Cook; sessions were chaired by Drs. Desmond King Hele, Stephen Pumfrey and John Henry. The help of all of these is gratefully acknowledged. The costs of the conference were borne by a generous grant from the Royal Society, and a reception was hosted by the British Library. Expenses relating to the background research for the conference and for the volume that has ensued were paid for by the Leverhulme Trust as part of an institutional grant to Birkbeck College for a project entitled 'Archives of the Scientific Revolution: the Case of Robert Boyle'. Others whose co-operation has been crucial include Mary Nixon and her staff at the Royal Society library, particularly Mary Sampson, who, in addition to her illustrated presentation, was also chiefly responsible for the exhibition that was put on in the library on the subject of the conference. A number of scholars generously responded to enquiries made in the course of planning the conference programme, and/or in connection with the preparation of the book. At Boydell and Brewer, Richard Barber has as usual proved a sympathetic and helpful publisher. The frontispiece is reproduced by kind permission of the President and Council of the Royal Society.

Notes on Contributors

MASSIMO BUCCIANTINI is Researcher in the History of Science at the University of Siena and collaborator at the Museum of the History of Science in Florence. He recently published the volume, *Contro Galileo: alle origini dell' "affaire"* (Florence, 1995).

CHRISTIANE DEMEULENAERE-DOUYÈRE has since 1989 been Archivist of the French Academy of Sciences, Paris. She has written or edited many books about scientific archives, most recently *Les archives personelles des scientifiques. Classement et conservation* (Paris, 1995) and *Histoire et mémoire de l'Académie des sciences. Guide de recherches* (Paris, 1996).

MORDECHAI FEINGOLD is Professor of Science Studies at Virginia Polytechnic Institute. His books include *The Mathematicians' Apprenticeship: Science, Universities and Society 1560–1640* (Cambridge, 1984) and (ed.) *Before Newton: the Life and Times of Isaac Barrow* (Cambridge, 1990). His monograph on the Oxford curriculum in the seventeenth century has just been published in volume 4 of the *History of Oxford University*. He is currently working on the history of the Royal Society.

MARK GREENGRASS is Professor of Modern History at the University of Sheffield. From 1988 to 1996 he was a Director of the Hartlib Papers Project, the principal product of which was the CD-Rom edition of *The Hartlib Papers* (Ann Arbor, 1995). He is currently working on an edition of Hartlib's *Ephemerides*.

FRANCES HARRIS is Curator in the Department of Manuscripts at the British Library and author of *A Passion for Government: a Life of Sarah, Duchess of Marlborough* (Oxford, 1991). She has joint responsibility for the Evelyn archive at the British Library, and is currently working on John Evelyn, with particular reference to his relationship with Margaret Godolphin.

ROBERT A. HATCH teaches in the Program for the History of Science, Technology and Medicine at the University of Florida. In addition to his *Collection Bouillau: an Inventory* (Philadelphia, 1982), he has written on the relations of science and learning in seventeenth-century France, with particular emphasis on the correspondence networks of Peiresc, Gassendi and Bouillau. He is currently writing a biography entitled *Bouillau*.

MICHAEL HUNTER is Professor of History at Birkbeck College, University of London. He has written or edited various books on late seventeenth-century English science and its milieu. With Edward B. Davis and Antonio Clericuzio, he is currently preparing a new edition of the *Works* and *Correspondence* of Robert Boyle.

ROB ILIFFE is Lecturer in the History of Science at Imperial College, London. He has published various articles on the cultural and social context of science in the early modern period. His edition of two of Newton's most significant theological treatises is to be published by the Voltaire Foundation in 1998.

DOMENICO BERTOLONI MELI teaches the history of science at Indiana University. He is the author of *Equivalence and Priority: Newton versus Leibniz* (Oxford, 1993, paperback 1996) and editor of *Marcello Malpighi, Anatomist and Physician* (Florence, 1997). He is currently working on the history of Italian medicine in the late seventeenth century.

JAMES O'HARA worked at various Dutch and German universities after completing his Ph.D. at the University of Manchester in 1979. In 1987 he joined the Leibniz Edition as an editor, and is now employed in editing Leibniz' mathematical, scientific and technical correspondence. He is joint editor of *Gottfried Wilhelm Leibniz: Sämtliche Schriften und Briefe*, third series, vol. 4, 1683–90.

DAVID STURDY is Senior Lecturer in History at the University of Ulster at Coleraine. He has published several books on European and French history, most recently *Science and Social Status: the Members of the Académie des Sciences 1666–1750* (Woodbridge, 1995). He is working on a study of the Abbé Jean-Paul Bignon (1662–1743) and the patronage of science.

JOELLA YODER is currently preparing a catalogue of the Codices Hugeniorum which will collate the manuscripts with the edition of Huygens' *Ouevres complètes*. She is the author of *Unrolling Time: Christiaan Huygens and the Mathematisation of Nature* (Cambridge, 1988).

Abbreviations

AN	Archives Nationales, Paris
BL	British Library, London
BN	Bibliothèque Nationale, Paris
Huygens, *Oeuvres complètes*	*Oeuvres complètes de Christiaan Huygens*, published under the auspices of the Holland Society of Sciences (22 vols., Amsterdam and The Hague, 1888–1950)
Newt*on* *Correspondence*	H.W. Turnbull, J.F. Scott, A.R. Hall and Laura Tilling (eds.), *The Correspondence of Isaac Newton* (7 vols., Cambridge, 1959–77)
OED	*Oxford English Dictionary*
Oldenburg, *Correspondence*	A.R. and M.B. Hall (eds.), *The Correspondence of Henry Oldenburg* (13 vols., Madison, Milwaukee and London, 1965–83)
RS	Royal Society
Whiteside, *Mathematical Papers*	D.T. Whiteside (ed.), *The Mathematical Papers of Isaac Newton* (8 vols., Cambridge, 1967–81)

1

Introduction

Michael Hunter

It is easy to take collections of manuscripts for granted as neutral repositories of data from which significant items can be individually garnered. This perception is encouraged by huge and wide-ranging deposits like the Additional Manuscripts of the British Library: such assemblages reflect a piecemeal process of prescient acquisition over many years, but by definition the result is a collection with an overall shape with no significance in itself. Smaller assemblages built up either by collectors or libraries in a similarly eclectic way show the same characteristics: such artefacts may reveal interesting information about the tastes of their creators, but, almost by definition, this is tangential to the individual significance of the items that they contain.

By contrast, archives which comprise the papers of individuals or institutions have a shape which is significant in itself. They are self-contained organisms, reflecting the purposes for which they were formed and the processes to which they have been subjected. This is nowhere truer than of the archives that have come down to us from the Scientific Revolution, which, despite recent controversy, remains the best title to encapsulate the epoch-making changes in the way in which the natural world was studied and conceptualised in the seventeenth century.[1] Many surviving manuscript assemblages were directly associated with the refashioning of natural knowledge that occurred at that time. These archives are thus themselves integral to the process of intellectual change: significance accrues to their individual components, their overall content, and the way in which the parts are organised to make up the whole; indeed, a sophisticated understanding of such matters should properly inform interpretation of their content. Equally important is what can be learned from the subsequent vicissitudes of these collections about changing attitudes towards the Scientific Revolution and its protagonists during the intervening centuries. They represent a crucial – if neglected – facet of our understanding of the period.

1 For recent discussions, see Roy Porter, 'The Scientific Revolution: a Spoke in the Wheel?', in R. Porter and M. Teich (eds.), *Revolution in History* (Cambridge, 1986), pp. 290–316; H.F. Cohen, *The Scientific Revolution: a Historiographical Enquiry* (Chicago, 1994); Steven Shapin, *The Scientific Revolution* (Chicago, 1996); John Henry, *The Scientific Revolution and the Origins of Modern Science* (Basingstoke, 1997).

The surviving materials fall into various categories. Perhaps most obviously, we have the more or less intact survival of the manuscripts of such intellectual giants as Galileo Galilei, Robert Boyle, Christiaan Huygens, G.W.Leibniz and Isaac Newton. These survive respectively as the Collezione Galileana at the Biblioteca Nazionale Centrale at Florence; the Boyle Papers at the Royal Society, London; the Codices Hugeniorum at Leiden University Library; and the Leibniz Archiv at the Niedersächsische Landesbibliothek, Hanover; while Newton's manuscripts survive partly in the University Library and at King's College, Cambridge, partly in the Jewish National and University Library, Jerusalem, and are partly scattered more widely.

Such collections have received a good deal of attention – if not always from as sophisticated a viewpoint as might have been desired – reflecting the traditional conception of intellectual change in the period as the work of heroic individuals. Such a view reached its peak in the late nineteenth century, inspiring the great editions of the writings of many such figures which remain at the core of the discipline; the relationship of such publications to relevant manuscript remains will be discussed elsewhere in this volume. But scarcely less is to be learned from similar – if usually smaller – surviving collections associated with less well-known figures who also played a key role in intellectual change in the period. In this volume, studies are included of two characteristic examples of these. One, representing the biological and medical sciences, is the archive of Marcello Malpighi, the bulk of it now in Bologna University Library. The other, the papers of Sir William Petty, in which science rubs shoulders with political economy, is a self-contained archive held for three centuries by Petty's family, which since 1992 has become a discrete component within the Additional Manuscripts at the British Library. These have been selected from a larger group of comparable collections belonging to figures associated more or less directly with the Scientific Revolution, some of which will be referred to later in this Introduction and some of which have been the subject of recent study: these include the papers of John Aubrey, John Evelyn, John Flamsteed and John Locke.[2]

If such deposits are significant largely because of what they reveal about the thought of the individual intellectuals who created them, equally important are the papers of 'intelligencers', men who promoted and serviced the networks of communication by which ideas and information were exchanged and spread at the time. Such archival remains give a more diffused view of intellectual change, more anchored in the cultural milieu of the day, of the kind that has come to the fore in recent years. Here, pride of place must go to the papers of Samuel Hartlib, now in Sheffield University Library, which have been the subject of a major programme

2 For bibliographical references, see below. For Locke, see also P. Long, *A Summary Catalogue of the Lovelace Collection of the Papers of John Locke in the Bodleian Library* (Oxford, 1959); id., 'The Mellon Donation of Additional Manuscripts of John Locke from the Lovelace Collection', *Bodleian Library Record*, 7 (1962–67), 185–93; John Milton, 'Locke at Oxford', in G.A.J. Rogers (ed.), *Locke's Philosophy: Content and Context* (Oxford, 1994), pp. 29–47; id.; 'Locke Manuscripts among the Shaftesbury Papers in the Public Record Office', *Locke Newsletter*, 27 (1996), 109–30; id., 'John Locke's Medical Notebooks', *Locke Newsletter*, forthcoming.

of exploitation and publication in recent years by the Hartlib Papers Project; this major holding is dealt with in chapter 3, below. But an equally significant – if less well-known – example is provided by the papers of the French savant, Ismaël Bouillau, the bulk of them now forming the Collection Bouillau in the Bibliothèque Nationale in Paris: these form the subject of chapter 4 of this book.

Overlapping in function are the manuscript deposits associated with the new, public institutions devoted to knowledge about the natural world that were founded during the period. Building on the work of pioneers like Hartlib and Bouillau, such bodies reflected a conviction that the advancement of learning required a greater element of co-operation and continuity than had existed hitherto. They laid the foundations for the institutionalised network of communication and the shared infrastructure for the evaluation of research findings which has characterised science ever since, and the establishment of which has often been seen as one of the key components of the Scientific Revolution.[3] A preoccupation with the creation and preservation of records was one of the key corollaries of such initiatives: indeed, as Mordechai Feingold points out in chapter 11, such functions were envisaged even in the various proposals for such bodies that preceded those actually set up. Here, the new scientific institutions are represented by the two most famous, the Royal Society of London, founded in 1660, and the Académie des Sciences of Paris, founded in 1666, though honourable mention should also be made of their precursor, the Accademia del Cimento, extensive documentation concerning which survives though it is not the subject of a discrete paper here.[4]

In fact, as the previous paragraphs reveal, the survival of archives directly associated with the formative stages of the Scientific Revolution is surprisingly good. This has long been reflected by the piecemeal use of material from these collections by historians of the ideas of the period. But it may be argued that insufficient attention has hitherto been given to the light that can be thrown on important issues in intellectual history by careful study of such archives as entities in their own right – their content, their organisation, and their treatment over the intervening centuries. It was to rectify this that a conference was held at the Royal Society, London, on 11–12 April 1996 entitled 'Archives of the Scientific Revolution'. Its object was to raise awareness of the issues involved by bringing together various scholars who had worked on related themes, and by encouraging others to think more fully than hitherto about the significance of archives as entities in their own right.

3 On the growing infrastructure for science, see esp. J.E. McLellan III, *Science Reorganised: Scientific Societies in the Eighteenth Century* (New York, 1985). See also J.L. Heilbron, *Elements of Early Modern Physics* (Berkeley and Los Angeles, 1982), ch. 2.

4 On the Accademia del Cimento, see esp. W.E.K. Middleton, *The Experimenters: a Study of the Accademia del Cimento* (Baltimore, 1971), and Paolo Galluzzi, 'L'Accademia del Cimento: "gusti" del Principe, filosofia e ideologia dell'esperimento', *Quaderni storici*, 48 (1981), 788–844. On archival aspects of the Royal Society and the Académie des Sciences, in addition to chs. 11 and 12 below, see esp. Eric Brian and Christiane Demeulenaere-Douyère (eds.), *Histoire et mémoire de l'Académie des sciences: Guide de recherches* (Paris, 1996) and Keith Moore and Mary Sampson, *A Guide to the Archives and Manuscripts of the Royal Society* (London, 1995).

The papers given at the conference – like the book which has resulted – deliberately concentrated on the heroic age of the Scientific Revolution from Galileo to Newton, which was also the age of the earliest scientific institutions. Within this, a range of subject matter was sought, as well as a variety of approaches. Thus, apart from the range from individual intellectuals to intelligencers and institutions that has already been indicated, readers will encounter various different types of investigation. These range from Robert Hatch's extraordinary detective work, which has allowed him to locate Bouillau's manuscripts in their scattered locations, to the vivid view of Boyle's disposition of his papers in his later years which is available from the extensive set of inventories of his unpublished writings that survive. In addition, the discussions that took place raised important broader issues concerning archives and their interpretation and exploitation. It is to these that this Introduction will be mainly devoted.

Of course, though the survival rate of scientific archives from the period c. 1630 to c. 1710 is good, it is far from perfect. Archives are fragile entities, vulnerable to a constant process of attrition, and it is worth sketching here the circumstances militating against their survival. To some extent, this was due to unavoidable catastrophe. Thus the manuscripts that Marcello Malpighi had built up in the early part of his career went up in flames early in 1684, when his house burnt down together with many of his possessions. This was witnessed by the English naturalist, Tancred Robinson, who reported how Malpighi 'comforted his Wife, and condol'd nothing but the Loss of his Papers, which are more lamented than the Alexandrian Library, or Bartholine's *Bibliothece* at Copenhagen' – thus alluding to earlier, comparable disasters.[5] In mid-seventeenth-century England, matters were made worse by the political and other upheavals that the country suffered. William Harvey's papers were lost when his rooms at Whitehall Palace were broken into when the Palace was deserted during the 1640s, which he told John Aubrey was 'the greatest crucifying to him that ever he had in all his life'; similarly, some of Jeremiah Horrox's papers were destroyed by soldiers during the Civil War, while others were lost in the Fire of London.[6]

But manuscripts were not only vulnerable to specific catastrophes. They were also affected by a more insidious process of attrition which means that the papers of many figures fail to survive, despite the fact that they are known to have existed at the time of their death. The traditional means by which manuscripts survived their owners was by descent through their family, and examples of archives which survived for long periods by such means and are dealt with in this volume are those of Malpighi, Newton and Petty. But this presented recurrent hazards, since some executors and descendants were more conscientious than others. One person who was particularly vociferous in deprecating the fate of manuscripts if left in the wrong hands was John Aubrey. Such matters bulk large in the word portraits of

5 H.B. Adelmann, *Marcello Malpighi and the Evolution of Embryology* (5 vols., Ithaca, 1966), i, 470.

6 John Aubrey, *Brief Lives*, ed. Andrew Clark (2 vols., Oxford, 1898), i, 303; A.B. Whatton, *Memoir of the Life and Labors of the Rev. Jeremiah Horrox* (London, [1859]), pp. 59f.

his contemporaries that comprise his famous *Brief Lives*, while his letters and other writings form a litany of complaints about the fate of archives if proper care was not taken of them – full of stories about papers used to wrap herrings, 'to stopp gunnes' or sacrificed to 'the tayler's sheeres'. Indeed a poignant survival among his manuscripts is a paper by the divine and natural philosopher, Seth Ward, which Aubrey has endorsed: 'I found this paper [which is his owne hand-writing], amongst his scatterd papers, which I rescued from being used by the Cooke, since his death; which was destinated with many other good papers & letters, to be put under Pies.'[7]

In fact, it is far from coincidental that the survival rate of archives from our period is as high as it is, for these years saw intense concern about manuscripts and their fate. In Aubrey's case, this reflects an acute sense of change stimulated by the traumatic events of the Civil War and its aftermath, a sense of change which itself partly inspired the foundation of the scientific institutions already referred to.[8] It is significant that, quite apart from the extent to which their function involved the creation of archives, contemporaries like Aubrey saw such bodies as having a role simply in terms of preserving manuscript material, even on topics somewhat tangential to their main area of activity.[9] It is also noteworthy how the Royal Society played a major role as a repository for the scientific papers of such overseas figures as Malpighi and Antoni van Leeuwenhoek.[10] In his own case, Aubrey's anxiety that his manuscripts should survive led him to 'make my owne hands my Executors' by consigning his literary remains to a 'publick Repository' in the form of the Ashmolean Museum, founded at Oxford by his friend, Elias Ashmole, in 1683, and his example was followed by the naturalist, Martin Lister.[11]

An equally explicit instance of conscious preservation is provided by Christiaan Huygens, who sought to ensure the survival of his papers by bequeathing them to the University of Leiden; Huygens' deposit, and his motives in making it, are discussed by Joella Yoder in chapter 6.[12] In a sense, a similar preservationist impulse underlay the activity of figures like Hartlib and Bouillau: as Robert Hatch shows in chapter 4, Bouillau seems to have seen his role as an intelligencer as extending to the preservation of the papers of former intellectuals, most notably Nicolas-Claude Fabri de Peiresc. It is perhaps not coincidental that from this period

7 Aubrey, *Brief Lives* (n. 6), passim; Michael Hunter, *John Aubrey and the Realm of Learning* (London, 1975), p. 65.

8 On this theme, see especially Michael Hunter, 'The Importance of Being Institutionalised', in id., *Establishing the New Science: the Experience of the Early Royal Society* (Woodbridge, 1989), pp. 1f. On Aubrey's sense of change, see id., *John Aubrey* (n. 7), pp. 162f.

9 For instance, see Aubrey's view that James Boevey's manuscripts should be preserved there: Aubrey, *Brief Lives* (n. 6), i, 114–15.

10 On Malpighi and the Royal Society, see Adelmann, *Marcello Malpighi* (n. 5), i, xviii and passim; on Leeuwenhoek, see Antoni van Leeuwenhoek, *All de brieven* (12 vols., Amsterdam, 1939–84, in progress), passim; L.C. Palm, 'Leeuwenhoek and other Dutch Correspondents of the Royal Society, *Notes and Records of the Royal Society*, 43 (1989), 191–206; E.G. Ruestow, *The Microscope in the Dutch Republic* (Cambridge, 1996), ch. 6.

11 Hunter, *John Aubrey* (n. 7), pp. 90, 92.

12 See also Joella Yoder, 'Christiaan Huygens' Great Treasure', *Tractrix*, 3 (1991), 1–13, esp. pp. 4–5.

dates the earliest attempt in Britain at a kind of National Register of Archives, in the form of Edward Bernard's great *Catalogue*, published in 1697: though obviously mainly concerned with collections of antiquarian significance, this also includes the holdings of such contemporaries with scientific interests as Robert Plot, Edward Tyson and Hans Sloane.[13]

Running in parallel, and interconnecting, with these mechanisms by which archives were preserved was the earliest impulse to their exploitation – in this case relating to efforts to record and celebrate the life and works of their originators, which can themselves be seen as a revealing facet of the new science in its formative stages.[14] The classic case of this is provided by the Galileo archive: though, as Massimo Bucciantini shows in chapter 2, Galileo had himself been aware of the significance of his manuscripts, in many respects the archive as an actual entity was created by Vicenzio Viviani in connection with his project for vindicating Galileo in the generation after the great man's death. This would have been achieved through a major collected edition of his writings and a life, though both proved abortive.

A not dissimilar state of affairs is found in the case of Robert Boyle. Boyle had initially taken it for granted that his family would be responsible for his literary remains, while ultimately, in 1769, his papers (or what was left of them) were bequeathed to the Royal Society.[15] But their history in the interim was subsumed by the same imperative that is found in Galileo's, namely of the great man's manuscripts being studied and utilised in connection with a project for writing his life. In this case, access to the archive was given to Boyle's putative biographer, the scholar William Wotton, who removed substantial quantities of material in connection with a life which never materialised. These documents remained separate from the rest of the archive until reunited in connection with a further project for celebrating Boyle through publishing his life and editing his works, this one associated with Thomas Birch and Henry Miles in the 1740s: Birch and Miles partially realised Wotton's goals by including in their edition of *The Works of Robert Boyle* not only correspondence printed from manuscript sources but also unpublished sections of Boyle's *The Christian Virtuoso*.[16]

On the other hand, though acting as an impulse to preserve and utilise archival materials, the effect of such projects could be mixed as far as the treatment of manuscripts was concerned. In Boyle's case, Wotton's removal of substantial quantities of his papers meant that they might never have rejoined the main body

13 Edward Bernard, *Catalogi Librorum Manuscriptorum Angliae et Hiberniae in unum collecti* (Oxford, 1697), pp. 73, 109–10, 251–5 and passim.

14 See Michael Hunter, 'Robert Boyle and the Dilemma of Biography in the Age of the Scientific Revolution', in Michael Shortland and Richard Yeo (eds.), *Telling Lives in Science* (Cambridge, 1996), pp. 115–37.

15 See id., *Letters and Papers of Robert Boyle: a Guide to the Manuscripts and Microfilm* (Bethesda, Md., 1992), pp. xif.; R.E.W. Maddison, 'A Summary of Former Accounts of the Life and Work of Robert Boyle', *Annals of Science*, 13 (1957), 90–108.

16 Hunter, 'Boyle and the Dilemma of Biography' (n. 14), esp. pp. 123–33; id., *Robert Boyle by Himself and his Friends* (London, 1994), pp. xxxvi–lxii.

of the archive but for Miles' assiduity in pursuing them. A similar state of affairs is revealed by Rob Iliffe in chapter 9, since abortive plans to publish theological writings by Newton in the eighteenth century led to their being split from the main body of the Newton archive; as a result, they were left for nearly a century in the hands of the family of their putative publisher prior to being presented to New College, Oxford, in 1872.

More insidiously, publication seems in some respects to have been seen as an alternative to preservation rather than a corollary of it: indeed, it may not be coincidental that the projects which had the most beneficial effect in preserving archives – those relating to Galileo and Boyle – were largely abortive. At the time, it seems to have been generally felt that manuscript materials, once published, were dispensible. This was evidently what Sir Robert Southwell had in mind when quoting to Petty the example of the divine and mathematician, Isaac Barrow, who left hardly any papers at his death, 'soe carefull had he beene to finish all he ever tooke in hand, either printing what he wrote, or leaving his thoughts and collections all ready for the Presse'.[17] The collecting of the published and unpublished writings of eminent scientists was a not uncommon feature of the period: examples include the edition of Horrox's writings produced by John Wallis, John Flamsteed and others in 1673, or the collected edition of John Wallis' own works produced in the 1690s, or the editions of writings by Robert Hooke published by Richard Waller and William Derham in the decades following his death.[18] In many such cases, the manuscript materials deployed in such editions have failed to survive. Equally revealing is the case of Descartes, whose correspondence was extensively published within a decade and a half of his death, and whose works were collected not long afterwards. For little effort seems to have been made to preserve his manuscripts, despite the fact that they survived near disaster when lost at sea en route from Sweden to France after Descartes' death and were retrieved and re-ordered by Descartes' follower, Claude Clerselier. Though some of the items that appear in the inventory of Descartes' papers made at Stockholm in February 1650 were seen in the late seventeenth century – not least by Leibniz, who made careful notes on them while in Paris in 1675–76, and by Descartes' biographer, Adrian Baillet, who deployed them in his *Vie de Descartes* (1691) – the originals neither of these nor of other items survive.[19]

The case of Boyle reveals a further, damaging effect of eighteenth-century biographical projects on the archive of their subject. Not only were many of Boyle's manuscript letters discarded after being published in Birch's and Miles' edition. In addition, there is evidence of a considerable degree of pruning of the

17 Quoted below, p. 89.

18 *Jeremiae Horroccii . . . opera posthuma* (London, 1673); John Wallis, *Opera mathematica* (3 vols., Oxford, 1693–99); Richard Waller (ed.), *The Posthumous Works of Robert Hooke* (London, 1705); William Derham (ed.), *Philosophical Experiments and Observations Of the late Eminent Dr Robert Hooke* (London, 1726).

19 Charles Adam and Paul Tannery (eds.), *Oeuvres de Descartes* (13 vols., Paris, 1897–1913; revised ed., Paris, 1964–76), x, 1–14, 171f., 205f. and passim. For details of earlier editions of Descartes, see ibid., i, passim. See also Stephen Gaukroger, *Descartes: an Intellectual Biography* (Oxford, 1995), p. 415.

archive, including the removal of letters deemed trivial or likely to detract from Boyle's reputation: these included family letters, begging letters, letters from alchemists and other 'enthusiasts', and the like. This is apparent from details given in inventories compiled by Miles of material of this kind that no longer survives, at least some of which one can deduce that he deliberately discarded since the items in question are endorsed 'No Worth'. The unfortunate effect of this has been to give the appearance that Boyle loftily transcended mundane affairs, and that he was less interested in pursuits such as alchemy than was really the case, from the historiographical effects of which we are only now escaping.[20] On the other hand, there was also a random element in the losses, since Miles' inventories reveal that material of which he can hardly have disapproved has also disappeared from the archive, presumably prior to its arrival at the Royal Society in 1769. This, too, has had an effect on the overall physiognomy of the archive as it has come down to us, and the result is that, even in a case like Boyle's where what survives is very extensive, conclusions based on this need to be qualified by the awareness that a massive and fascinating body of material is lost.[21]

These late seventeenth- and early eighteenth-century developments might be seen as the first phase of the posthumous fortunes of the archives dealt with here. But interest in the manuscript remains of the leading lights of the Scientific Revolution was reinforced in the late eighteenth and nineteenth centuries by a growing impulse to historical and antiquarian study, in some cases (as in Italy and the Netherlands) reinforced by nationalist overtones. Thus it was in the 1770s that the papers of John Flamsteed – which have recently received significant scholarly attention – were acquired by the Board of Longitude and deposited at the Royal Observatory, Greenwich (they were first tabulated and exploited by Francis Baily in the 1830s).[22] In other cases, archives that had been preserved in the families of their originators, but with a carelessness that had led to increasing harm being done to them, now received proper attention. Frances Harris points out how it was in the 1780s that a serious interest was first taken in the Petty archive, while, in Italy, Domenico Bertoloni Meli recounts how Malpighi's manuscripts were rediscovered in the early nineteenth century by Gaetano Atti, who expressed amazement at the neglected state of the riches he found, thereafter cataloguing them and arranging for their deposit at the University of Bologna.

More significant still was the history of the Galileo archive, which came to light in a neglected condition in the late eighteenth century, but which was taken into public ownership in 1818 and subjected to extensive attention in the 1830s: indeed, it is from this episode that its current arrangement largely dates, including its division into classes devoted to Galileo, his predecessors, contemporaries, disci-

20 See Hunter, 'Boyle and the Dilemma of Biography' (n. 14), esp. pp. 131–2.

21 An annotated text of Miles' inventories is in preparation by Michael Hunter and Lawrence Principe.

22 See Adam Perkins, 'The Flamsteed Papers in the Archives of the Royal Greenwich Observatory', in Frances Willmoth (ed), *Flamsteed's Stars* (Woodbridge, 1998), pp. 217–24, on pp. 218–21. See also the other contributions to that volume and E.G. Forbes, Lesley Murdin and Frances Willmoth (eds.), *The correspondence of John Flamsteed* (2 vols., Bristol, 1995–7) [in progress].

ples, and 'posteriori', thus summarising the entire history of Italian science.[23] Such impulses continued through the nineteenth century, resulting in similar attention being paid to the Leibniz archive; in the examination of the Newton archive and the gift of what was deemed important in it to the University of Cambridge in 1872; and in the beginning of the great project for editing Huygens' papers that forms the focus of Joella Yoder's treatment in chapter 6, a project comparable to Antonio Favaro's 'national edition' of Galileo (1890–1909), which, as Bucciantini shows, grew out of the archival study that had occurred earlier in the century.

As earlier, the effects of such attention were sometimes mixed. Perhaps worst was the ancillary effect of such preservationist impulses in the diaspora associated with the impact of collectors, especially of autographs. A case in point was William Upcott, who undoubtedly did piecemeal damage to many groupings with which he came into contact in the course of building up the magpie-like accumulation that was housed at his residence, 'Autograph Cottage', in Upper Street, Islington. Much light on the activities both of men like Upcott, and of collectors on a larger scale such as Sir Thomas Phillipps, was shed by the studies of the late A.N.L. Munby, whose documentation of the effect of their interference on collections that had remained intact up to that time deserves to be better known by scholars.[24] Moreover, even if the more significant English archives dealt with here were largely immune from the attentions of such men, the same was not true in Europe, where the booming market in such manuscript material inspired the scholar and librarian, Guglielmo Libri, to a sustained series of thefts and forgeries, the effect of which on French and Italian scientific archives is dealt with in both chapters 4 and 12 below.[25]

On the whole, the work of scholars in the nineteenth century reflected the increasingly professional standards in historical study which characterised that period. But inevitably this was predicated on contemporary attitudes towards science and its history, and some of the side-effects of the scholarly attention that archives received at this time were unfortunate. Most striking is the splitting up of the Newton archive, so that only the 'serious scientific' material went into public custody, the remainder – deemed scientifically unimportant – being left with the family. For a while this component remained intact, if neglected, but the Victorian demarcation provided the mandate for the Sotheby's sale of 1936, at which it was scattered to the winds at bargain prices (it is perhaps worth recalling that though John Maynard Keynes and Abraham Yahuda were respectively able to assemble 130 and thirty-nine lots, this was out of a total of 332).[26] Moreover, the attention that archives *did* receive sometimes reflected criteria which were inappropriate to

23 See below, chapter 2.

24 See A.N.L. Munby, *The Cult of the Autograph Letter in England* (London, 1962), p. 18 and passim; id., *Phillipps Studies* (5 vols., Cambridge, 1951–60; reissued, London, 1971).

25 For a recent study of Libri, see P. Alessandra Maccioni Ruju and Marco Mostert, *The Life and Times of Guglielmo Libri (1802–69)* (Hilversum, 1995).

26 See below, chapter 9; see also P.E. Spargo, 'Sotheby's, Keynes and Yahuda – the 1936 Sale of Newton's Manuscripts', in P.M. Harman and Alan Shapiro (eds.), *The Investigation of Difficult Things* (Cambridge, 1992), pp. 115–34.

the collection: cases in point include the rather artificial classificatory system applied to the Boyle Papers when they were bound up for the first time in the 1850s,[27] and above all the saga recounted by Joella Yoder in chapter 6, where the integrity of the archive was quite ruthlessly subordinated to the perceived priorities of the edition of Huygens' *Oeuvres complètes* that was then put in hand.

Such vicissitudes are significant in themselves, and a number of contributors to this volume dwell on them at length. Much is to be learnt from them about the evolving reputations of the figures whose archives are involved, and about changing attitudes to the history of science. In addition, in most cases we have to live with decisions made in the past as to the selection preserved and/or its method of classification, a further incentive to understand them better. Only rarely – as with recent acquisitions by a repository like the British Library – has it proved feasible to undo such a classification and to re-order the manuscripts in a manner more appropriate to their original state. Moreover, even when such a reorganisation is undertaken, it is almost as dependent on an understanding of the phases through which an archive has passed as on an appreciation of how it originated.[28]

Whether in conjunction with such opportunities for re-ordering or not, however, the really tantalising goal is to get beyond the changing fortunes that archives have experienced – intrinsically interesting as these may be – to try to reconstruct their purpose and content in the hands of their originators. This is more feasible in some cases than others, and some of the authors in this volume have not been able to penetrate beyond the changes to the archive that separate us from its original state. In the case of both Leibniz and Newton, regrettably, the evidence on such matters seems particularly elusive. Even here, however, it is worth fastening on such clues as survive, in the latter case, the evidence of Newton's working practices which Rob Iliffe notes, or the suggestion that he ruthlessly censored his papers before he died.[29] But in other cases, the peeling away of strata of superimposed orderings – or the survival of evidence such as listings, endorsements and other clues to the original state of the archive – makes it possible to tackle truly fascinating questions relating to the manuscripts in the hands of their original owners, in terms of content, organisation and use. Such issues are addressed in various of the essays that follow.

Yet here, too, problems arise, which are not unique to archives from this period but are shared by all such assemblages. One issue is, when is an archive deemed to be 'complete'? Is the crucial point the moment of its originator's death, or might some earlier date be equally or more appropriate? In fact, the impulses to manipulate an archive for biographical purposes, sketched above in relation to the generations following an author's death, often began in the latter years of an author's life. This subject has been studied most fully in connection with nineteenth-century literary authors by Michael Millgate in his book *Testamentary Acts* (1992). Millgate there illustrates the extent to which figures like Tennyson and

27 Hunter, *Letters and Papers* (n. 15), pp. xv–xvi, xxvi–vii.

28 See Frances Harris' paper given at the conference, 'Archives of Seventeenth to Twentieth Centuries', St Hugh's College, Oxford, 29 June 1995.

29 See below, p. 152; Whiteside, *Mathematical Papers*, i, x.

Hardy destroyed documents and otherwise manipulated their literary legacy to ensure that the image of them that was transmitted to posterity was consistent with their own view of themselves in their later years, a process that their executors often only continued.[30]

Similar evidence may be found among the figures studied here, whether or not this was effective, as with the items which, although they still survive among Boyle's papers, are endorsed 'These Papers are without fayle to be burn't.'[31] Similarly, Bertoloni Meli indicates how Malpighi seems to have been solicitous as to which parts of his intellectual legacy were preserved and published, while, though others, such as Petty, showed less concern, in his case this role was taken on his behalf by his friend, Sir Robert Southwell.[32] Similar considerations also entered when manuscripts were bequeathed to public institutions, as with Aubrey's bequest to the Ashmolean which has already been referred to, for he felt the need to indulge in an element of self-censorship: though he left the bulk of his manuscript writings to the museum, there were some which went elsewhere, apparently because he considered them 'too light', not seeing them as 'fitt to leave to the Critiques of the University'.[33]

Such instances might be taken to suggest that collections like those dealt with here experienced a critical moment in their history when they were transferred from being the working papers of a living intellectual to being the 'archive' of a dead icon. But such a distinction cannot be sustained. For one thing, the concept 'archive' was applied by Petty and Southwell to his collection during his lifetime, and even if others were more likely to talk about their 'papers', this does not reflect a fundamental difference of conception.[34] Secondly, the selection and reorganisation that might occur in an author's later years in the expectation of death was often the last stage in a longer process of selection and organisation that spread throughout his life. Indeed, almost by definition, archives are constantly evolving during their owners' lifetime, and one of the challenges in a book like this is to understand the decisions involved. No-one keeps every scrap of paper that passes through their hands, and the manner in which those whose archives are studied here selected items for retention is a key topic for us to comprehend.

Such partial retention may have varied according to date within a man's life. In some cases there is evidence of a disproportionate amount of material surviving from a specific period. An example is that of Petty, as Frances Harris points out, where the survival of papers from his later years is particularly profuse, evidently due to his exceptional output at that time. Interestingly, as she points out, he had reclassified the archive prior to this intense period of activity. In other cases, there is evidence of authors sharing the sense that print was an alternative to the retention

30 Michael Millgate, *Testamentary Acts: Browning, Tennyson, James, Hardy* (Oxford, 1992).

31 Royal Society Boyle Papers 26, fol. 162. Attention was first drawn to this item, Boyle's early essay 'Of the Atomical Philosophy', in R.S. Westfall, 'Unpublished Boyle Papers on Scientific Method', *Annals of Science*, 12 (1956), 63–73, 103–117, on pp. 110–11.

32 See below, chapters 5, 7.

33 Hunter, *John Aubrey* (n. 7), p. 91n.

34 See below, chapter 5.

of manuscripts that we have already encountered. Thus men who retained unpublished writings from earlier periods in their lives often seem to have seen the manuscripts of works that had appeared in print as dispensible: this was true of both Huygens and Boyle, though in Boyle's case he retained material from works that were partially published, in some cases retaining published material because it was intermingled with unpublished material from the same work.

Some archives are remarkable for the extensive series of working papers that they contain, which long precede publication on the part of the author. Here, perhaps the classic case is that of Newton. No significant portion of Newton's mathematical research appeared in print until 1687, though some items had previously been circulated informally. As a result, the main clues to the earlier evolution of his thought are the papers that he preserved from the early 1660s onwards.[35] Indeed, as Iliffe remarks in chapter 9, the published editions of Newton's mathematical working papers, notebooks and the like – which survived only in manuscript until our own time – now form the chief source for the study of Newton's thought.

To some extent, an understanding of such matters may depend on making sense of things that were obvious to the originator of the archive, such as the date of documents within it. An important tool in making judgments on such matters is the analysis of handwriting. In the case of Newton, D.T. Whiteside was able to deploy the skill that he developed in dating his subject's handwriting to great effect in allocating material to different phases in his life.[36] With Boyle, a similar state of affairs exists in that though he wrote very little himself, it has proved possible to identify the distinct handwriting of a number of amanuenses who worked for him at different stages in his career: since the turnover of such amanuenses was relatively rapid, this provides if anything an even more secure dating method (though obviously, as with an author's own handwriting, it only provides a terminus ante quem, and allowance has to be made for earlier items being recopied).[37] On the whole, the potential of watermarks has so far proved limited, especially in view of the cumbersome procedures required to study them.[38]

Related to the question of the different phases of an individual's creative life represented in an archive is the range of types of material to be found in it, and how these were categorised and organised. It goes almost without saying that no two archives are the same, and that the very make-up of the archive is revealing of the intellectual personality of its creator. Here various issues arise, which the authors of different chapters are able to comment on in greater or lesser detail. One is the extent to which an author's papers reflect the interests displayed in his published writings, and how far we see a different, more private world: a particu-

[35] Whiteside, *Mathematical Papers*, i, xv and passim.

[36] Whiteside, *Mathematical Papers*, esp. i, p. xi n. 2.

[37] See Hunter, *Letters and Papers* (n. 15), pp. xxviii–xxxix.

[38] See A.E. Shapiro, 'Beyond the Dating Game: Watermark Clusters and the Composition of Newton's *Opticks*', in P.M. Harman and A.E. Shapiro (eds.), *The Investigation of Difficult Things* (Cambridge, 1992), pp. 181–227.

larly notorious case of this is presented by the alchemical and theological materials that make up such a substantial portion of Newton's manuscript holdings. Equally crucial is the issue of how papers were demarcated. In Boyle's case, a clear division recurs throughout his life between his theological and philosophical writings. Beyond that, it is possible to observe a distinction on his part between a core of compositions on which he was working at any particular time, and a penumbra comprising older, discarded papers, and more tangential items such as notes, copies of papers by others, and compendia of data.

Both in Boyle's and other cases, the most immediately interesting items in an archive may be those relating directly to an author's intellectual evolution; more will be said about these in relation to the issue of editing material from archives later in this Introduction. But the more miscellaneous materials with which these are juxtaposed are hardly less significant, even if a different approach may be required to make sense of them. Such items include notes and copies of documents by others, representing a formative stage in an author's ideas. One typical component found in many archives of this period are commonplace books, records of the reading of their originator. In the case of natural philosophers, this category may overlap with notes on observations and experiments, especially since many prided themselves on studying 'things' rather than 'words' – though the extent to which this was a rhetorical pose is revealed by the large libraries that they often also owned.[39]

Commonplace books – which scientists were as likely to keep as others – have been the subject of a good deal of study in relation to the sixteenth and seventeenth centuries as a whole; Harold Love has commented on the 'culture of transcription' which characterised the period, noting the evident satisfaction that laymen derived in copying out and classifying data in this way.[40] Though the rationale of such practices may have been changing by the period on which this book is focused, they still retained considerable vitality.[41] A good example is provided by the extensive records of this kind to be found among the Evelyn Papers, now in the British Library, while equally interesting is the method of commonplacing advocated in an article published in the *Bibliothèque universelle* in 1686 by John Locke, which sufficiently impressed Samuel Pepys for him to keep a copy of it among his papers.[42] Whether Pepys actually used the method is unclear – in fact, it appears

39 Hunter, *John Aubrey* (n. 7), p. 40; id., *Science and Society in Restoration England* (Cambridge, 1981, reprinted Aldershot, 1992), p. 151.

40 Harold Love, *Scribal Publication in Seventeenth-century England* (Oxford, 1993), p. 200 and passim, esp. ch. 5. For studies of commonplace books see J.M. Lechner, *Renaissance Concepts of the Common-places* (New York, 1962); W.H. Sherman, *John Dee: the Politics of Reading and Writing in the English Renaissance* (Amherst, Mass., 1995), esp. chs. 3–4; Ann Moss, *Printed Commonplace-Books and the Structuring of Renaissance Thought* (Oxford, 1996).

41 Moss, op. cit., chs. 8–9; Richard Yeo, 'Ephraim Chambers's *Cyclopaedia* (1728) and the Tradition of Commonplaces', *Journal of the History of Ideas*, 57 (1996), 157–75.

42 G.G. Meynell, 'John Locke's Method of Commonplacing', *The Seventeenth Century*, 8 (1993), 245–67; Samuel Pepys, *Private Correspondence and Miscellaneous Papers*, ed. J.R. Tanner (2 vols., London, 1926), i, 146; see also above, n. 2. For Evelyn's commonplace books, see *John Evelyn in the British Library* (London, 1995), pp. 36–8.

that Locke himself was far from systematic in deploying it – and surviving materials of this kind cry out for more intensive study. Yet this is one of the key areas where one can actually see archives at work, observing how data was collected, processed and retrieved through different systems of notation. It is all the more significant in view of the interest of intellectuals of the day like Robert Hooke in finding means by which data might be as readily as possible managed and deployed.[43]

Commonplaces came from books, and this brings us to libraries and their relationship to the archives that form the subject of the current volume. Certainly, there is an interrelationship between the two. In fact, archives and libraries often overlapped, and some of the holdings dealt with here interpenetrated with the books of the same individual. In Newton's case, some of the earliest evidence concerning the evolution of his mathematical ideas comes from notes that he made from printed texts, some of which survive in their own right.[44] In other instances, we are fortunate to have the complete survival of an author's library as well as his manuscripts, perhaps the best example being that of Leibniz, where the two have survived together ever since his own time, as James O'Hara indicates in chapter 10. Ideally, it should be possible to study the interrelationship of the two, observing a process of collocation of information from the printed page to manuscript digests and drafts and back onto the printed page again: such opportunities may be provided by the Evelyn collection, now that his archive and a selection of his books are once again housed under the same roof.[45]

Here, libraries have been eschewed, but it is worth noting a further potential area of overlap between them and archives, namely in terms of how books were organised. Indeed, there has been a good deal of interest in the often elaborate systems of classification devised for libraries in this period, perhaps particularly in France.[46] Less attention has been paid to the question of how these related to the way in which books were actually shelved: Samuel Pepys, for instance, stuck resolutely to a classification by size, even if attempting a more analytical scheme in his library catalogue, while a recent study of Evelyn's library suggests that the manner in which his books were arranged was less sophisticated than the system to which he aspired.[47] Archives also overlap with another classic phenomenon of the intellectual culture of the period, the cabinet of curiosities. Here, manuscripts might overlap with more tactile holdings, perhaps the classic case being the assemblage of Sir Hans Sloane, which combined objects and ancient and more

43 Hooke, *Posthumous Works* (n. 18), pp. 63–5.

44 Whiteside, *Mathematical Papers*, i, passim; John Harrison, *The Library of Isaac Newton* (Cambridge, 1978), appendix C and passim.

45 See *John Evelyn in the British Library* (n. 42), pp. 84–5.

46 See, for instance, Henri Martin, 'Classements et conjunctures', in *Histoire de l'edition francaise, vol. 1, Le livre conquerant* (Paris, 1983), pp. 429f.; Claude Jolly, 'Naissance de la "science" des bibliothèques', in *Histoire des bibliothèques françaises, vol. 2, Les bibliothèques sous l'Ancien Regime* (Paris, 1988), pp. 381f.

47 Robert Latham (gen. ed.), *Catalogue of the Pepys Library at Magdalene College, Cambridge* (7 vols., Woodbridge, 1978–94), esp. vol. 7, *Facsimile of Pepys' Catalogue*, ed. David McKitterick, 1991, pp. xi–xxx; Guy de la Bédoyère, 'John Evelyn's Library Catalogue', *The Book Collector*, 43 (1994), 529–48.

recent manuscripts.[48] Such collections also raised issues of tabulation and classification which bear some analogies to those arising in the case of larger assemblages of manuscripts.

Libraries and cabinets bring us to the overlap between the private and public domain, since a number of both effected a transition from the former to the latter. Not only did Sloane's collection become the basis of the British Museum; the same had been true earlier of the Royal Society's repository, based on the collection of Robert Hubert, and of the Ashmolean Museum, which similarly built on the assemblage of the Tradescants and of Elias Ashmole himself.[49] Similarly, private libraries often formed the basis of early public ones, as with Narcissus Marsh's library in Dublin, for instance – assembled out of various collections, including Marsh's own and that of Edward Stillingfleet – or the library founded by Thomas Plume at Maldon, Essex.[50] Yet, in the course of this, the intentions and functions of such collections changed. This is particularly clear in relation to a museum like that of the Royal Society, where the whole point of such a transition was to make it possible for the original private collection to form the core of a larger, more comprehensive and more utilitarian one.

This brings us to the archives of early scientific institutions dealt with in chapters 11 and 12 of this book. To some extent, these could be seen as the institutional analogues to an individual's working papers. A case in point is provided by Christiane Demeulenaere-Douyère's and David Sturdy's analysis of the *procès-verbaux* of the Académie des Sciences in its early years, and of the papers of the academy's laboratory, in which the scientific policy of the institution can be seen evolving. Beyond this, however, the archive of a public body arguably served rather different functions from those of private individuals, paralleling the change in the nature of libraries and museums when they effected the private-public transition. One aspect of this was a slightly legalistic one, of the use of public deposit as a means of registering proprietorship on the part of inventors and others. This was particularly notable in the early years of the Royal Society, when its facilities were used for such purposes by various figures, including Boyle;[51]

48 Arthur MacGregor (ed.), *Sir Hans Sloane, Collector, Scientist, Antiquary* (London, 1994). See also P.M. Jones, 'A Preliminary Check-list of Sir Hans Sloane's Catalogues', and M.A.E. Nickson, 'Hans Sloane, Book Collector and Cataloguer, 1682–98', *British Library Journal*, 14 (1988), 38–51, 52–89. On museums, see also Oliver Impey and Arthur MacGregor (eds.), *The Origin of Museums: the Cabinet of Curiosities in Sixteenth- and Seventeenth-century Europe* (Oxford, 1985); Krzysztof Pomian, *Collectors and Curiosities: Paris and Venice, 1500–1800* (Eng. trans., Cambridge, 1990); Paula Findlen, *Possessing Nature: Museums, Collecting and Scientific Culture in Early Modern Italy* (Berkeley and Los Angeles, 1994).

49 See Hunter, *Establishing the New Science* (n. 8), ch. 4; Arthur MacGregor (ed.), *Tradescant's Rarities: Essays on the Foundation of the Ashmolean Museum, 1683* (Oxford, 1983).

50 Muriel McCarthy, *All Graduates and Gentlemen: Marsh's Library* (Dublin, 1980), esp. ch. 2.; S.G. Deed, comp., *Catalogue of the Plume Library at Maldon, Essex* (Maldon, 1959); W.J. Petchey, *The Intentions of Thomas Plume* (Maldon, 1985).

51 For a review of this issue, see Michael Hunter, 'The Reluctant Philanthropist: Robert Boyle and the "Communication of Secrets and Receits in Physic" ', in O.P. Grell and Andrew Cunningham (eds.), *Religio Medici: Medicine and Religion in Seventeenth-century England* (Aldershot, 1996), pp. 247–72, on pp. 263–4.

later, as Feingold reminds us in chapter 11, the archive was put to a related use in the precise claims about intellectual transactions made in the notorious *Commercium Epistolicum* (1712). Indeed, as Demeulenaere-Douyère and Sturdy point out, in serving this function, the archives of early scientific institutions were fulfilling a purpose analogous to those of more traditional public archival holdings.

In addition, of course, such repositories served a broader recording function, in line with the Baconian aspirations characteristic of the science of the day. The accumulation and preservation of data was thus part and parcel of the ambition of such bodies 'to scrutinise the whole of Nature and to investigate its activity and powers by means of observations and experiments', in the words of Henry Oldenburg, the Royal Society's first secretary.[52] Yet here too they tended to have a different character from a 'private' archive, being the receptacles of discrete items, separately 'registered', hence being more bureaucratic in character than their private equivalent and at least in theory more accessible to a range of interested parties who might wish to draw on the data so recorded.

In addition, from the outset correspondence bulked large in the holdings of such bodies. Indeed, correspondence is one of the chief areas of overlap between 'private' and 'public' archives, frequently being itself seen as a kind of 'public' possession once a letter had left the hands of its author.[53] It is hence appropriate to consider this specific component of such holdings here. In some respects, this is the most familiar type of manuscript material surviving from the period, yet its interpretation raises various problems, a number of them – perhaps paradoxically – relating to archives. In fact, a clear definitional demarcation needs to be established at the outset. Archives are quintessentially discrete groups of manuscripts, often of a single owner; this gives their actual content – and their omissions – a significance in its own right along the lines already sketched. A correspondence, on the other hand, is not so delimited. A correspondence may be defined as the epistolary interchange between two or more people, and is usually described in relation to one of these: in other words, the correspondence (say) of Newton comprises the letters between Newton and all of his various correspondents. By definition, such a correspondence will survive in more than one place. Moreover, this can be extended to types of material which, even if not strictly 'letters' in the normal sense, were treated by both their originators and their recipients in a similar way: an example, as Bertoloni Meli shows in chapter 7, are the *consulti* of Malpighi.

Reconstructing a correspondence cuts across archival boundaries. In other words, a correspondence is an entity which has a kind of ideal existence, which an edition realises in a way that has no direct archival equivalent. The whole point of a correspondence is to illustrate networks rather than individual deposits, and any one archive will have a kind of refracted equivalent in others, mediated

52 Oldenburg to Van Dam, 25 Jan. 1663, in *Oldenburg*, ii, 13–14.

53 On this point, see Frances Willmoth's introduction to *The Correspondence of John Flamsteed* (n. 22), ii, xliii–iiv.

through the correspondence divided between them. This is the point which Mark Greengrass imaginatively illustrates in relation to the Hartlib Papers in chapter 3.

Yet, even here, we return to the inescapable reality of archives, since, whatever the ambition of the editor of a correspondence to transcend them, it is actually through archives that correspondence survives. Letters are normally to be found in the holdings of their recipient, even if some figures were more careful than others about retaining such material. In the seventeenth century one hardly ever, if at all, encounters the practice which became common in the late eighteenth and nineteenth centuries, of letters being returned to their author either by their recipients or by his or her executors, so that they could be preserved and published as part of an entire correspondence. In this period, one is usually dependent on the survival of the archives of recipients, and, since this is differential, it needs to be taken into account in interpreting the material: thus in the case of the correspondence of J.A. Comenius and John Dury, as Greengrass points out, both are heavily weighted towards their disproportionate survival among the Hartlib Papers.

Such difficulties do not arise in relation to authors who routinely kept copies of their letters, meaning that the actual letters sent were duplicated in the holdings of their originator. But such cases where copies of letters survive in the sender's own archive can in fact be more problematic than may appear at first sight. A particularly clear example of this is provided by John Evelyn, who was assiduous in keeping such items. In some cases, there is reason to believe that letters retained in this way may never have been sent. In others, the version kept differs from that dispatched, sometimes being retrospectively embellished and misdated. The result is to present an editor's nightmare, with which various scholars are currently struggling.[54] In other cases, the problem is less acute, but here too the survival of a letter in the archive of its originator always needs to be handled with caution, not least since it is not clear, unless there is corroborative evidence, that the letter was in fact ever sent.

If even correspondence has archival complexities which have to be taken into account in publishing it, these are still greater in the case of the other kinds of material contained in holdings of the kind dealt with in this book. It is therefore worth briefly considering these here. Certain archives may be appropriate for complete publication, particularly when an entire archive – or an entire section of it – comprises correspondence and associated documentation: here, the more general criteria relating to the publication of epistolary exchanges come into play. An example of this might be the Boyle Letters (itself, in fact, a creation of eighteenth- and nineteenth-century date which has no direct mandate from Boyle himself but makes sense as a separate entity in this way).[55] The Hartlib Papers can similarly be seen as a kind of extended correspondence, to which, as Mark Greengrass argues in chapter 3, Hartlib's 'Ephemerides' may be seen as a guide, with ancillary documents being keyed to the letters to which they relate.

54 See esp. Guy de la Bédoyère, *Particular Friends: the Correspondence of Samuel Pepys and John Evelyn* (Woodbridge, 1997).

55 See Hunter, *Letters and Papers* (n. 15), p. xix.

On the other hand, more complex challenges are presented by the typically diffuse archives dealt with in the pages that follow. As we have seen, these often range from correspondence, through original working papers and drafts by the author in question, to notes and copies from the work of others. The exploitation of such assemblages requires a series of properly informed decisions which it is worth sketching here, so as to stimulate fuller consideration of such matters and to avoid an uncritical attitude towards archives, a kind of fetishism of citing manuscripts almost for the sake of it.[56]

As a matter of record, it is clearly highly desirable that such holdings should be made widely available through some form of reproduction. Of these, the easiest to achieve is microfilm. Not only does this increase the availability of the material in question to interested parties (and reduce wear on the originals); it also means that subsequent decisions which involve selective exploitation can be readily checked by any who have reservations about the criteria underlying them. Almost equally important is cataloguing, an undervalued art which forms the basis of all subsequent work on an archive, and which, though currently unfashionable, is in some respects more fundamental than editing. At its most basic, the outcome of this would be a simple descriptive calendar. But at its most sophisticated it would effectively represent an analysis of the archive, based on a full understanding of the subject matter of the collection and of the handwritings encountered in it. It would thus tabulate what material was in holograph, showing signs of composition, and what items were scribal copies; it would seek as far as possible to date the material; it would indicate what was duplicated, either within the archive or outside it; and it would link notes to their sources and drafts to the works to which they related. Such analysis is crucial if the student of an archive is to execute an appropriate editorial function, for it is on the basis of such informed tabulation of the material in question that all subsequent decisions relating to it will be made.

Once an archive has been tabulated and analysed in this way, it is possible to adopt a sophisticated attitude towards each discrete item within it, rather than to treat the holding as an undifferentiated whole (though all such decisions will, of course, be informed by a sense of the overall physiognomy of the collection in hand). Indeed, since exploitation in the form of citation or publication normally focuses on specific items, rather than on the entire archive, such understanding of the exact status of each component part dealt with in this way is particularly crucial. Here, it is worth establishing the concept of a 'document', as the unit around which analysis should focus, a discrete item which – if published – should be published as a whole or not at all. Once an archive begins to be approached in this way, it will soon transpire that there are varying grades of significance, and hence of publishability, among its components, and it is worth considering these here.

We may start with the kind of 'penumbra' which survives in many archives – copies of works by others, notes and commonplace books. Clearly, there is no point

[56] For further discussion of related issues, see Michael Hunter, 'How to Edit a Seventeenth-century Manuscript: Principles and Practice', *The Seventeenth Century*, 10 (1995), 277–310, esp. pp. 284f.

in transcribing multiple copies of the same text, or scribal transcripts of texts readily available in print: here, comparison and tabulation of variants is more appropriate. Moreover, we here return to the role of careful analysis, often based on the study of handwriting and on knowledge of sources other than the archive itself, since such skills may be required even in deciding whether a document is an original composition or a copy. A notorious case of this is provided by the Newtonian 'Clavis', once thought to be a composition by Newton himself but now known to be merely his copy of a work by someone else, almost certainly George Starkey.[57] Indeed, the issue of how Newton's disparate manuscripts on alchemical and related topics should be published has proved far more problematic than has been the case with his mathematical manuscripts, not least because of the rather different status of much of the material involved, much of it involving notes and copies of the work of others rather than original compositions.

Depending on their nature, even preliminary drafts for published works which survive in an archive may be better dealt with by tabulation, collation and analysis than by complete transcription, which simply leaves individual readers the task of collation which an editor might have undertaken on their behalf once and for all. An example is provided by the draft material that survives for some of Boyle's writings, including his *Free Enquiry into the Vulgarly Receiv'd Notion of Nature* (1686). Much of this overlaps fairly closely with the published text, although significant conclusions can be drawn from it in terms both of the overall evolution of the work, and the way in which Boyle modified his piecemeal ideas in specific passages.[58] On the other hand, in other cases drafts may be so different from the published version that full publication is appropriate: discretion is required in assessing each instance and dealing with it on its merits.

What about more finished, but unpublished, writings that survive in an archive? In many ways these are the most exciting component of an archive, but this also raises the stakes in assessing their significance, to an extent that is not always appreciated. What is crucial is that their presence in the archive should not be taken for granted, since their survival in unpublished form itself requires explanation and analysis. Indeed, publication without such exegesis may be misleading in itself. Here, interesting questions arise relating to the intentions of the archive's originator, and the degree of circulation to which he aspired for his writings. The presence of an item in unpublished form in an archive may be easy to explain in the case of insubstantial or incomplete working papers, but a different kind of explanation is required in the case of complete works that never saw the light of day. It is especially important to understand why a work remained in manuscript if it seems likely that this was linked to external constraints on the author. Here, it is essential that the archive is not viewed in isolation, but is understood in terms of the social milieu in which the author operated. In Petty's case, as Frances Harris points out, he clearly sought a limited circulation for the brief papers which he

57 See William Newman, 'Newton's *Clavis* as Starkey's *Key*', *Isis*, 78 (1987), 564–74.

58 Michael Hunter and Edward B. Davis, 'The Making of Robert Boyle's *Free Enquiry into the Vulgarly Receiv'd Notion of Nature* (1686)', *Early Science and Medicine*, 1 (1996), 204–71.

increasingly committed himself to writing as his life progressed, while Rob Iliffe points out the extent to which Newton, too, released papers to a select circle. Boyle's archive contains some significant items which are in virtually finished form but which he thought better of publishing, for reasons ranging from intellectual rigour on his part to anxieties about professonal demarcation.[59] Basically, anything that is in an archive is there for some reason, and it is important to understand what this is.

Of course, the resolution of such issues will be specific to each individual case. The most difficult of these is probably that of Newton, whose generally acknowledged brilliance is matched only by his intellectual convolution, and the problems relating to whom appear to be particularly intractable, as Rob Iliffe indicates in chapter 9. What should be apparent, however, is that such decisions on the status and significance of items within an archive must be informed by a full understanding of the kind of considerations concerning evolution, survival, and the like that have been canvassed in this Introduction. In part, what is needed is an almost forensic examination of the working of the archive as it existed in the hands of its originator. Equally important is a knowledge of the history of its subsequent vicissitudes, which is significant both for what it reflects about the way in which science and its practitioners have been perceived over the intervening period, and for its implications for the assemblage's component parts. In sum, it could be said that there is a kind of undiscovered 'archaeology' of archives: it is to enhancing understanding of such holdings and their descent, and to illustrating the conclusions that may be drawn from this, that this volume is devoted.

Acknowledgments

I am grateful to various of the contributors to this volume for their helpful comments on a draft of this Introduction, and to John Cottingham, Stephen Gaukroger, Giles Mandelbrote and Frances Willmoth for advice on specific points.

[59] See Michael Hunter, 'Boyle versus the Galenists: a Suppressed Critique of Seventeenth-century Medical Practice and its Significance', *Medical History*, 41 (1997), 322–61; id., 'Reluctant Philanthropist' (n. 51), passim; Hunter and Davis, 'Making of Boyle's *Free Enquiry*' (n. 58), esp. pp. 236–42, 265–6.

2

Celebration and Conservation: the Galilean Collection of the National Library of Florence

Massimo Bucciantini

I

The history of the Galilean Collection was told in great detail just over a century ago by Antonio Favaro, the eminent Galileo scholar who edited the twenty volumes of the National Edition of the Italian scientist's works.[1] The long study entitled *Documenti inediti per la storia dei manoscritti galileiani nella Biblioteca Nazionale di Firenze* that Favaro published in 1885 is still today an irreplaceable source for anyone wishing to become acquainted with the progressive formation of one of the most important archives of the Scientific Revolution.[2] In those same years, working with Alarico Carli, Favaro also produced an analytical catalogue of the whole collection which, although unfortunately unpublished, is still a useful point of reference; it is to be found in the indexes and catalogues section of the manuscripts room of the Central National Library in Florence.[3]

The project for a published catalogue of the Galilean Collection was finally realised, albeit slowly, from the 1950s onwards, thanks to the solitary and selfless commitment of Angiolo Procissi. Procissi, a professional mathematician and passionate devotee of the history of mathematics, published the first volume, dealing with files 1–99, in 1959; the second (files 100–154) appeared thirty years later in 1985; while the third and fourth volumes (files 155–258) were published posthumously in 1994.[4] Paolo Galluzzi and Maurizio Torrini subsequently contributed further to pinpointing and clarifying the complex events leading to the

I wish to thank Ailsa Wood for the English version of this paper.

1 G. Galilei, *Opere*, ed. A. Favaro (20 vols., Florence, 1890–1909).

2 Favaro's article was published in *Bullettino di bibliografia e di storia delle scienze matematiche e fisiche*, 18 (1885), 1–112, 151–230. For further information on the Galilean Collection see: L. Andreani, 'I manoscritti di Galileo e della sua Scuola nella Biblioteca Nazionale Centrale di Firenze', *La Bibliofilia*, 11 (1909), 44–61; D. Fava, *La Biblioteca Nazionale Centrale di Firenze* (Milano, 1939), pp. 108–13: A. Procissi, 'I manoscritti superstiti dell'Accademia del Cimento', in *Celebrazioni dell'Accademia del Cimento nel tricentenario della fondazione* (Pisa, 1958), pp. 25–34.

3 Central National Library, Florence, Manuscripts Room, A. Favaro-A. Carli, *Inventario dei manoscritti galileiani*, 3 vols., signature: Catal. 14a.

4 A. Procissi (ed.), *La Collezione Galileiana della Biblioteca Nazionale di Firenze* (Rome, 1959–94): vol. 1: *Anteriori e Galileo*, 1959; vol. 2: *Contemporanei e Discepoli*, 1985; vol. 3 (tomi 1–3), I. Truci e M.

birth of this archive: the vicissitudes suffered by the manuscripts of Evangelista Torricelli, Vincenzo Renieri and other followers of Galileo, which were later united in the Galilean Collection, have been exhaustively described in the introduction to the first volume of the Correspondence of Galileo's Followers, published in 1975.[5]

It is not my intention to retrace all the phases of this story, characterised – like so many other stories of archival transmission – by dispersion and fortuitous rediscoveries, by unspeakable negligence and passionate and generous acts of self-denial. Instead I propose to focus on the extraordinary wealth to be found, even today, within the Collection, both for the specific study of Galileo and, more generally, for research relating to the Scientific Revolution as a whole.

The Galilean Collection is composed of more than 50,000 papers, distributed in 347 volumes. These contain all the works and autograph letters of Galileo – which have now mostly been edited – together with writings and documents relating to his friends and correspondents, including some of the greatest scientists and philosophers of the period, such as Kepler, Mersenne, Gassendi and Peiresc. The Collection also includes the extraordinarily complete documentation of the activities of the Accademia del Cimento, in particular the records of that body's experiments and its writings and correspondence. Lastly, we have most of the manuscripts and letters of Galileo's principal followers – Benedetto Castelli, Bonaventura Cavalieri, Evangelista Torricelli, Vincenzio Viviani – the main representatives of the 'Galilean school'. As well as the wealth of correspondence between these leading figures of seventeenth-century science and philosophy and the most important members of the European intellectual community, the origins and development of various problems and topics of great interest can be followed through the Collection: from debates about the vacuum, and about recurrent observations of comets and other astronomical phenomena, to discussions arising from numerous experiments on heat, light and magnetic phenomena.

II

It is always difficult to determine *the starting point* of the history of an archive. All such reconstructions risk seeming arbitrary or – more often – elusive and ambiguous. Indeed it is no coincidence that studies of the foundation of archives

Zangheri (eds.): *I Discepoli (parte II, V. Viviani)*, 1994. A few months before his death in 1987, Procissi managed to finish cataloguing the whole Collection. The revision and correction of the fourth and last volume, concerning files 259–347, is being carried out by Isabella Truci of the Central National Library of Florence. On Angiolo Procissi see G. Sestini, 'Angiolo Procissi (1908–87)', *Torricelliana*, 38 (1987), 191–2.

5 P. Galluzzi and M. Torrini (eds.), *Le Opere dei Discepoli di Galileo. Carteggio* (2 vols., Florence, 1975–84), i, xvii–xlii. On the Renieri manuscripts preserved in the Galilean Collection see also A. Favaro, *Amici e corrispondenti di Galileo Galilei. XII. Vincenzo Renieri* (Venezia, 1905), pp. 157–62, 190–5 (now in A. Favaro, *Amici e corrispondenti di Galileo*, edited and with introduction by P. Galluzzi, (3 vols., Florence, 1983), i, 517–22, 550–5).

and collections of documents frequently appear vague and imprecise compared to the scrupulous and at times excessively detailed attention given to the examination of the various stages of their subsequent institutionalisation. I do not, however, believe that this differential state of affairs is attributable to a shortage of information or a lack of research. Often the cause of this is linked to the fact that, if one wishes to arrive at the origins of a historical event like the formation of an archive or a collection of books and manuscripts, one has to consider matters which are intertwined with more general questions and seem to have little to do to with the specific problem in hand.

When was the original core of the Galilean Collection formed? Or rather, when did the Italian scientist's papers and documents begin to assume the appearance and characteristics of an archive to be preserved and protected for the whole scientific community, rather than a private collection? It is difficult to avoid replying to these questions by saying that the founder of the Galilean Collection, the person who was first fully aware of the role and importance those papers would have in the birth of a new way of looking at nature and the universe, was Galileo himself. It is, in fact, the care Galileo took of himself and his image – above all starting from the publication of the *Sidereus Nuncius* (1610) – which most strikes us on each reading of his works and correspondence. So much is this so that we are persuaded that the first person to consciously work towards the construction of the Galileo myth in all its iconographic and documentary aspects was Galileo himself; not at the end of his very troubled life but long before, that is, from the years of his greatest success thanks to the unanimous recognition of his extraordinary discoveries about the skies.

In general, it is only in the second half of the seventeenth century that we see the dispersal of important collections frequently being prevented or at least limited by the foundation and diffusion of scientific academies; by the ever more institutional nature of libraries and scientific and natural historical collections; and, last but not least, by a new attitude and greater sensitivity towards documentation. The significance of this is illustrated by the general contrast between the state of affairs in Italy in the early and late seventeenth century. If we look at the archives of scientists and philosophers who lived in the first half of the century, with the exception of the extraordinarily rich General Archive of the Company of Jesus[6] and the documentation preserved in the Galilean Collection of the National Library at Florence, very little remains of the collections of manuscripts and correspondence of scientists like Giovanni Antonio Magini (1555–1617), Giovambattista Baliani (1582–1666), Benedetto Castelli (1578–1643), and Bonaventura Cavalieri (1598–1647). Therefore it is to Galileo himself that the principal credit is due for the preservation of his manuscripts, or rather of his early writings; of important works on motion which date back to before his Pisan period; and then, from the

6 See information in U. Baldini, 'Su alcune fonti archivistiche per la storia della scienza', in R. Cremante and W. Tega (eds.), *Scienza e letteratura nella cultura italiana del Settecento* (Bologna, 1984), pp. 567–74.

Paduan period, of the wealth of scientific and family correspondence, including many letters written by him of which he kept a copy.

It can easily be assumed that the well-known events surrounding Galileo's trial and sentence had a serious impact on the Collection's history. Suor Maria Celeste, Galileo's first born daughter, wrote on 13 July 1633:

> Sig. Geri was here one morning while we thought you were in trouble and along with Sig. Aggiunti he carried out that task in your house, which he says he told you about, which still seems right and necessary to me in order to avoid all the unfortunate incidents that could have happened. Therefore I had no reason to deny him the key and the permission to do so, as I saw that he was extremely keen to protect your interests.[7]

Galileo was travelling to Florence. He had left Rome about two weeks before, where on 22 June in the large hall of the Dominican Monastery of Santa Maria sopra la Minerva he had been condemned 'ut haereticus suspectus de vehemente' and compelled to recant. Fearing lest the Inquisitor of Florence should take possession of Galileo's papers, Suor Maria Celeste along with the trusted Geri Bocchineri and Niccolò Aggiunti thus decided to bring about the 'disappearance' of the more compromising documents and writings.

After Galileo's death on 8 January 1642, most of his papers and writings were left in the hands of his son Vincenzo, later passing to his mother, Sestilia Bocchineri, in 1649 and subsequently inherited by Vincenzio Viviani. It is interesting that barely two days after Galileo's death, Vincenzo, in a letter to Viviani, then absent from Florence, exhorted him to take care of a chest in which his father's manuscripts were kept safe:

> And should you hear that anyone for any reason wants it or wishes to have it removed from where it is, you must prevent them and hide it by having it taken elsewhere, and if necessary to the convent, and giving it to my sister.[8]

These are isolated testimonies, but they say a great deal about the climate of suspicion and coercion surrounding the final years of the Italian scientist's life. Nor should we be surprised by the fact that in Galileo's will no mention is made of his library, his unedited writings or his manuscripts.

III

The peculiarity of the Galilean Collection does not so much lie in its history during Galileo's own lifetime, as in the events that occurred immediately after his death. It was at this point that the plans for the formation and enrichment of the Collection merged into a single, great project for the redemption and reinstatement of the

7 Letter quoted in Favaro, 'Documenti inediti' (n. 2), p. 5.
8 V. Galilei to V. Viviani, 11 January 1642, quoted in Favaro, 'Documenti inediti' (n. 2), pp. 66–7.

heretical scientist. This was the beginning of a process of adulation which was to continue in subsequent centuries to produce the many myths of Galileo which have characterised the civil and cultural life of our country, finally culminating, at the end of the nineteenth century, in the construction of that everlasting monument of Italian culture, the publication known as the National Edition, edited by Antonio Favaro.

The key figure in these early developments was Vincenzio Viviani, a significant natural philosopher in his own right and Galileo's most devoted disciple. When Galileo died, Viviani was not yet twenty years old. However, it is mainly thanks to his tireless and devoted work in collecting Galileo's writings that the majority of the Galileo manuscripts have survived for posterity. Viviani dedicated much of his life to a single purpose: the integral publication of the works of the Maestro. The celebration of the idealised image – for this is what it was – of Galileo was not, however, simply an expression of the profound attachment and gratitude felt by a loyal and devoted follower, which Viviani undoubtedly was. In this case, just as in other periods of Italian history, the revocation of Galileo assumed specific forms and clearly defined meanings.

The principal goal that Viviani set himself was to have all Galileo's works 'reprinted in folio form in the greatest fullness and magnificence in two columns for the two languages, the one Tuscan, in which the author wrote, and the other Latin . . . With the addition of a great many of his writings that had been lost, which with great difficulty I have collected from various places'.[9] Viviani initially cultivated the Bolognese printer Carlo Manolessi in order to achieve the publication of Galileo's works: Manolessi's edition (1655–56), however, included neither the condemned *Dialogue* nor the *Copernican Letters* to Benedetto Castelli and Grand Duchess Christina, and both Prince Leopold de' Medici and Viviani were deeply disappointed and dissatisfied with it.[10] In the edition that they projected, the *Dialogue* and the *Copernican Letters* would naturally be present among Galileo's published works; and among his unedited works the writings on longitude would be featured. Moreover, the many letters that Viviani collected from all over Europe would aid him in the writing of the *Life of Galileo*, a vast and detailed biography of Galileo which would be included in this first full edition of his works.

Little more than forty years had passed since Galileo, with the extraordinary astronomical innovations of his *Sidereus Nuncius*, had unexpectedly claimed the attention of what we would now call the international scientific community. And yet centuries seemed to have passed since then. The reinstatement of Galileo through the integral publication of his works would have unequivocally signalled an inversion of opinion, ending a situation which for too long had been sterile and frustrating. The dedication and determination with which Viviani applied himself to this editorial task must be read within this framework: the four-volume folio edition, put together with the greatest fullness and magnificence and including

9 V. Viviani to M. Thevenot, 6 May 1661, quoted in Favaro, 'Documenti inediti' (n. 2), p. 159.
10 See Favaro, 'Documenti inediti' (n. 2), pp. 18–19, 80–92.

suspect and prohibited material, was intended to be remembered as one of the great commemorative events of the century.

The monumental *Life of Galileo*, written in honour of the great mathematician and philosopher by his follower, which was to have prefaced the edition, would have set an appropriate seal on this outstanding venture. Indeed, Viviani had begun to prepare for this task some months before the Maestro's death, writing to Borelli and Renieri asking for writings and documents.[11] Thereafter, as the many letters sent by Viviani to France, Holland and Germany bear witness, the search for Galilean papers and documents was carried out with extreme care and on a huge scale. The idea met with a favourable reception, so much so that numerous correspondents sent Viviani all the letters and writings in their possession.

But at the root of this request there also lay a secret and unconfessable worry: that of verifying the possible existence of compromising documents which, if revealed, could have rebounded against Galileo himself, thus placing at risk the plan for scientific and religious reconciliation that Viviani and the Grand Duke set in motion. Any commentaries on lesser known episodes of the trial and sentence, or the longitude affair undertaken with the Protestant Dutch States, could, if published, present a serious threat to the positive outcome of the negotiations and indeed contribute to the reinforcement of Galileo's image as a heterodox and heretical scientist. Thus the news of the imminent publication in Amsterdam of Paolo Sarpi's correspondence, which contained a number of letters from Galileo, was enough to cause serious alarm and concern among his Italian friends and followers. 'I was inwardly upset at this news', wrote Viviani on 24 July 1673 to Magalotti, who was in Holland,

> because it immediately came into my mind that, if it were true, a great deal of material evidence would be made available to the sworn enemies of Galileo – as you know there are hoards of them – to suspect him of that which he certainly was not... I know that, if I were in those parts, I would go on purpose to Amsterdam to identify those letters myself and, having seen them, whatever they were, not only would I quietly try any trick or means to prevent the publication of these Galilean letters, but I would also try to remove the originals and also the copies, at any monetary cost; and if they were already in print, I would myself pay for these papers, so that they should not exist in any other's memory.[12]

The examination of Galilean manuscripts promoted all over Europe by Viviani and Prince Leopold was thus becoming a control and prevention operation of any

[11] See Favaro, 'Documenti inediti' (n. 2), pp. 67–70.

[12] L. Magalotti, *Delle lettere famigliari* (Florence, 1769), p. 45. On this interesting document see P. Galluzzi, 'I sepolcri di Galileo: le spoglie 'vive' di un eroe della scienza', in L. Berti (ed.), *Il Pantheon di Santa Croce a Firenze* (Florence, 1993), pp. 150–1; and the introduction by L. Borsetto to V. Viviani, *Vita di Galileo. Con appendice di testi e di documenti* (Bergamo, 1992), p. 40. Not wholly dissimilar to Viviani's abortive life of Galileo is the project of Birch's life of Boyle as true philosopher and Christian. On the choice and the omissions of the material for Birch's *Life* see now M. Hunter, 'Robert Boyle and the dilemma of biography in the age of the Scientific Revolution', in M. Shortland and R. Yeo (eds.), *Telling Lives in Science: Essays on Scientific Biography* (Cambridge, 1996), pp. 115–37, on pp. 128–34.

parallel unauthorised editorial initiative which might have disturbed, and in some cases even proved contrary to, the ambitious Tuscan project concerned with consecrating and commemorating Galileo's true image as a Catholic scientist and man of faith.

Viviani's *Life*, moreover, was intended to help to put an end to the many indiscretions and malicious insinuations that had long accumulated around the figure and work of Galileo. 'The world wants to know the truth about that case', wrote Sir Robert Southwell to Viviani from London,[13] highlighting the expectations which surrounded Viviani's project during the decades when it was in progress. In fact, as is well-known, the *Life* never saw the light of day, just as the hoped-for Galilean edition was not published in Viviani's lifetime.[14] Even if things had turned out differently, however, it is clear that many questions regarding the events surrounding the sentence and trial would still have remained unanswered.[15]

The endless negotiations and public duties which afflicted Viviani during his life, and about which he often complained in letters to friends, were only apparently responsible for the failure. The real causes were to be found elsewhere. The wearing and ever uncertain negotiations with influential members of the Society of Jesus, the objective of which was to obtain the lifting of the publication ban on the *Dialogue* – an indispensable action for the eventual publication of the complete works – were swept aside when, at the close of the century, the quarrel between modernists and traditionalists flared up again:

> All Rome is up in arms against the mathematicians and physico-mathematicians. They have formed, and are forming, special congregations dependent on the Holy Office and the general prohibition of all authors of modern physics is discussed in front of the Pope, and long lists of them are drawn up; and among these, at the top of the list, are Galileo, Gassendi and Descartes for being extremely ruinous to the Republic of Letters and to the sincerity of the Faith.[16]

13 Favaro, 'Documenti inediti' (n. 2), p. 35. On the success of Galileo in England, see M. Feingold, 'Galileo in England: the first phase', in P. Galluzzi (ed.), *Novità celesti e crisi del sapere. Atti del convegno internazionale di studi galileiani* (Florence, 1984), pp. 411–20. On the relationships between England and Italy in the second half of the seventeenth century see now S. Gómez López, 'The Royal Society and Post-Galilean science in Italy', *Notes and Records of the Royal Society*, 51 (1997), 35–44.

14 On Viviani and, in particular, his *Life of Galileo*, see A. Favaro, 'Vincenzio Viviani e la sua "Vita di Galileo" ', *Atti del Reale Istituto veneto di scienze, lettere ed arti*, 62 (1902–3), 683–703. Further information in Viviani, *Vita di Galileo* (n. 12), pp. 11–50; M. Segre, 'Viviani's Life of Galileo', *Isis*, 80 (1989), 207–31.

15 This is deduced from some of the Viviani-Ricci correspondence: 'Pages 33 and 34: I would moderate those terms of persecution and such – observed Michelangelo Ricci – when speaking of Father Grassi and other Jesuits and imitators. Page 35: It is necessary to limit those praises of the book about the two systems [the *Dialogue Concerning the Two Chief World Systems*] to the many natural problems solved in it, so that it cannot be said that Copernican opinions are being defended and approved': Favaro, 'Documenti inediti' (n. 2), p. 165. And also: 'I consider Your Lordship's advice prudent, to use few words when discussing the details of the misfortune that befell Galileo over the *Dialogues*, and to expatiate further on other astronomical and geometrical matters, his meaningful experiences and wise reflections': M. Ricci to V. Viviani, 13 March 1668, ibid., p. 179.

16 V. Viviani to M. Malpighi, 3 February 1693, quoted in C. Dollo, 'Inediti per l'epistolario malpighiano', *Rivista di storia della filosofia*, 39 (1984), 545. Viviani received this information from a Roman correspondent, the Jesuit Antonio Baldigiani: see Baldigiani's letter to Viviani of 25 January 1693,

Any possibility of a compromise solution, or for the new and more advanced outcome hoped for by Viviani, was again crushed beneath the weight of the violent polemic which broke out against the Neapolitan democrats and atheists.

IV

On Viviani's death in 1703, his vast book collection was donated to the Library of the Hospital of Santa Maria Nuova, according to instructions in his will, in which, however, no information was given regarding the Galileo manuscripts. One reason for what may seem a strange omission is provided by Targioni Tozzetti in his monumental work entitled *Notices of the Rise of Physical Science in Tuscany* (1780). Targioni expressed his acknowledgment and gratitude for the humble and difficult task carried out by Viviani:

> Which of us in the last century, or the present one, during the reign of His Serene Highness Grand Duke Cosimo III, could have published a Life of Galileo, and praised his deeds without running great risks or at least having it tell against him? There was above all his devoted follower Vincenzio Viviani, who collected with great care and at any price everything he could of Galileo's writings, mathematical instruments, his Lincei Academy ring, his correspondence and his books?

But Targioni also remembered how Viviani, forced to work in secret and in justifiable fear of being searched, kept all the writings of Galileo and his followers and correspondents hidden in a wheat hole 'and there they remain to this day'.[17]

The subsequent stages of this interesting and complicated story involve the Florentine nobleman, Giovan Battista Nelli, whose prescient retrieval of the main body of the manuscripts once possessed by Viviani[18] was apparently followed by

published by A. Favaro, 'Miscellanea galileiana inedita', *Memorie del Reale Istituto veneto di scienze, lettere ed arti'*, 22 (1882), 850. On this matter, see M. Torrini, *Dopo Galileo. Una polemica scientifica (1684–1711)* (Florence, 1979), pp. 7–40. On the Neapolitan atheists, see L. Osbat, *L'inquisizione a Napoli. Il processo degli ateisti 1688–1697* (Rome, 1974).

17 G. Targioni Tozzetti, *Notizie degli aggrandimenti delle scienze fisiche accaduti in Toscana nel corso di anni LX del secolo XVII*, vol. 1 (Florence, 1780), pp. 121, 124.

18 Light on this is shed by Nelli's brief will, dated 18 July 1750 and published by Favaro: 'I bought from Messrs. Carlo and Angelo Panzanini [the Panzanini family had inherited the famous collection on the death of Viviani in 1703] a large number of manuscripts of Galileo Galilei, Vincenzio Viviani, Evangelista Torricelli, Gio. Alfonso Borelli, together with a great abundance of mathematical instruments which belonged to Vincenzio Viviani, all for eighty-eight *scudi*. My great fortune in obtaining such a wonderful treasure so cheaply came about through the ignorance of the people selling it, who were not aware of the value of those manuscripts and instruments. Most of these writings were hidden in a hole for storage of wheat in the "Casa dei Cartelloni" belonging to me [which formerly belonged to Viviani, was at that time the home of the Panzanini brothers and would later belong to Nelli himself]. Galileo's writings were collected by Vincenzio Viviani and a portion of them began to be sold by Messrs. Panzanini to the pork butchers [to be used for wrapping cured meat and sausages], just as the original writings of Vincenzio Viviani were sold, however I tried to buy those that he still had': Favaro, 'Documenti inediti' (n. 2), p. 201.

an unsuccessful attempt to sell the collection to the French government.[19] On Nelli's death in 1793, however, the tortuous history of the Galilean manuscripts reached a turning point. Nelli's will left no doubts whatsoever:

> The abovementioned gentleman [i.e. Nelli] has remembered that he possesses a well-stocked library with many of the famous Galileo's manuscripts . . . and orders that, should his children and heirs wish to sell this library, before entering into negotiations with anyone they should let His Royal Highness know so that he may buy it for the public libraries of this city. Thus the works of worthy men collected by the bequeather with research, care and expense, may not be lost.[20]

In October 1818, after long negotiations, Grand Duke Ferdinand III purchased the whole Nelli Collection for 1046 zechins, including the manuscripts of Galileo, Torricelli, Viviani and the Accademia del Cimento. The archive was deposited in the Regia Biblioteca Palatina, and thus the core of the actual Galilean Collection assumed for the first time the institutional and public nature that any great collection of manuscripts deserves.

An equally crucial period for the Galileo legend was the definitive ordering of the Collection in the second decade of the nineteenth century. The celebration of the great Tuscan scientist now began to acquire a new and rather different meaning: he was transformed into a glorious national symbol around which the very identity of Italian philosophy and science would be rediscovered. With Viviani, as we have seen, the Galileo polemic was characterised by reinstatement and redemption: the recognition of Galileo as an obedient and devout scientist was the price to be paid in order to overcome the isolated and subordinate position in which Italian science found itself. In the nineteenth century, on the other hand, the image of Galileo as uncontested father of the new philosophy, founder of the experimental method, took on a new aspect, assuming strong civil and political connotations: Galileo became the symbol of national identity, forerunner of that cultural and political unity which was indispensable in order to compete for civil and political supremacy among the nations.[21]

The definitive reordering of the Collection was an important part of this plan.

19 This much can be deduced from a letter dated 5th April 1760, sent by Pierre Jean Grosley (1718–85) to Chrétien Guillaume de Malesherbes (1721–94), in which it seems clear that Nelli had offered to sell the family archive, including the Galilean manuscripts and Machiavelli's writings, to the French Government: 'De l'instant où vous avez bien voulu interesser à la negotiation pour les Ouvrages de Galilée, je n'ai plus été inquiet du succés. Je ne doute pas que Mr. Nelli ne tienne la parole qu'il vient de vous donner aussi solennellment. Il serait bon de lui donner aussi quelques coups d'éperon au sujet des Considérations Politiques de Machiavel et sul le Commentaire de Cesar. Il m'a dit les avoir parmi le mss. de son cabinet. Il est vrai qu'il ne me les a pas fait voir. Ce morceau seroit pour les Politiques ce que les Oeuvres de Galilée seront pour les *hauts-savants*': Favaro, 'Documenti inediti' (n. 2), p. 205. Unfortunately the letter, which was kept in the Biblioteca Mediceo-Laurenziana in Florence, is a solitary document and does not allow us to understand the reasons for this offer.

20 Quoted in Favaro, 'Documenti inediti' (n. 2), p. 58. For further information see ibid., pp. 58–60, 206–30.

21 For the influence of the Galileo myth on Italian cultural and civil history, see E. Garin, *La cultura italiana tra '800 e '900* (Bari, 1962), pp. 29–79; P. Galluzzi, 'La scienza e la tecnica', in G. Moggi-P. Roggi (eds.), *Firenze, 1815–1945: un bilancio storiografico* (Florence, 1990), pp. 337–57.

When Vincenzio Antinori, a Florentine scientist and director of the Regio Institute and Physics Museum,[22] presented his plan for the arrangement of the Collection to Grand Duke Leopold II, it was immediately clear that this was an excellent means to combine preservation and commemoration. The five sections into which the Collection is still divided today served symbolically to indicate the main stream of the development of science:

Central National Library, Florence

General Prospect of the Galilean Collection (c. 1830)

I.	Predecessors	vols.	10
II.	Galileo's Manuscripts	vols.	89
III.	Contemporaries	vols.	11
IV.	Followers	vols.	148
V.	Successors	vols.	49

From halfway through the sixteenth century until the eighteenth, the glorious age of the Scientific Revolution was divided by Antinori into four eras (the Predecessors, Contemporaries, Followers and Successors of Galileo), dependent on and subordinate to a fifth era, that of Galileo himself. Could there be a better way of reminding Italian and foreign scholars who read those precious manuscripts of the role played by the Tuscan scientist and his school in the founding of modern civilisation?

The reorganisation of the archive was accompanied by the issue of a new, ostensibly definitive edition of Galileo's works, the so-called Grand Duke's edition, founded under the aegis of Grand Duke Leopold II and published in sixteen volumes between 1842 and 1856. Conceived in the atmosphere of euphoria which characterised the Galilean celebrations at the Third Congress of Italian Scientists which took place in Florence in 1841,[23] the 8,000 odd pages of the Florentine edition marked a notable progress compared to all previous Galilean editions.[24] It included works never included in any previous edition of Galileo, and gathered together a significantly greater number of letters than had been known up until that time. However the edition was deeply affected by the violent polemics which broke out between Vincenzo Antinori and Eugenio Albèri (1807–78) concerning the discovery in the Biblioteca Palatina of unedited observations

22 On Antinori see G. De Caro in *Dizionario Biografico degli Italiani*, vol. 3 (Rome, 1961), *ad vocem.*

23 The Congress took place in Florence in 1841. On cultural and philosophical life in those years, and in particular on activities carried out by the Italian Scientists Congresses, cf. E. Garin, 'Il Congresso pisano degli Scienziati italiani del 1839', *Giornale critico della filosofia italiana*, 69 (1990), fasc. 3, 281–2.

24 And particularly, compared to both the thirteen volume edition of the Italian Classics published in 1808–11, and that of the Biblioteca Enciclopedica Italiana published, also in Milan, in 1823, by Nicolò Bettoni. On Eugenio Albèri see F. Fonzi in *Dizionario Biografico degli Italiani*, vol. 3 (Rome, 1961), *ad vocem.*

by Galileo on the satellites of Jupiter; immediately after the publication of the second volume, these polemics ended in Antinori and Pietro Bigazzi resigning from their duties as advisor and literary assistant to the work.[25]

It was perhaps partly for these reasons that, barely twenty-five years after the conclusion of Albèri's edition, the project for a new edition was begun, Antonio Favaro's National Edition, which was formally approved by the Minister of Education, Michele Coppino, and sealed by royal decree dated 20 February 1887. Favaro many times exposed the unreliability of the earlier edition due to arbitrary interpolations and manipulations; indeed in retrospect his judgment may seem unduly harsh. Without doubt, however, Favaro's National Edition represented the greatest tribute to the memory of Galileo. The first volume came out in 1890, and publication continued without interruption, concluding with the twentieth volume in 1909.

Perhaps the principal significance of that extraordinary and crucial work, which still remains today, is the stimulus that it provides to study and understand Galileo through his works and unedited writings. As Favaro was proudly to declare on several occasions, the new and really complete edition of Galileo would not reproduce a single line of which the autograph original, or at least a contemporary copy, existed without a thorough collation being made.[26] The twenty volumes of the National Edition, together with the bottomless pit of information represented by the publication of 'Friends and Correspondents', 'Adversaria' and 'Galilean Fragments', form a seldom-equalled monument to philological and historical learning.[27]

V

Favaro's editorial feat provided an appropriate climax to the long and fruitful process of Galilean commemoration during the nineteenth century. Reissued between 1929 and 1939 during the fascist regime, it was inevitably endowed with meanings and purposes that differed vastly from its original intentions: as with many other propagandistic undertakings of that period, it formed part of a political campaign in open support of the regime, directed towards the vindication of the historical primacy of Italian science, and making Galileo's genius a source both

25 The Antinori-Albèri polemic stirred up a great deal of scandal. The accusations and counter-accusations hurled by the protagonists at one another were made public with a series of printed statements and letters. For a complete list see *Bibliografia galileiana (1568–1895)*, collected and illustrated by A. Carli, ed. A. Favaro (Rome, 1896), pp. 197–202.

26 A. Favaro, *Per la edizione nazionale delle opere di Galileo Galilei. Esposizione e disegno* (Firenze, 1888), p. 32. On Favaro see M. Bucciantini in *Dizionario Biografico degli Italiani*, vol. 45 (Rome, 1995), *ad vocem*.

27 Some of Favaro's main works on Galileo have recently been reprinted. As well as the monographs of the *Amici e corrispondenti di Galileo* series (n. 5), see the following anastatic reprintings edited by Lucia Rossetti and Maria Laura Soppelsa: A. Favaro, *Adversaria galilaeiana: serie I–VII* (Trieste, 1992); A. Favaro, *Scampoli galileiani* (2 vols., Trieste, 1992).

of national pride and of exaltation of the regime: the same was to happen with Leonardo da Vinci and Alessandro Volta.[28]

Already in 1942 an attempt had been made, on the initiative of Pietro Pagnini, to resume the Galilean project with the republication of the *Essays on Natural Experiences*.[29] But it was above all at the beginning of the 1970s that interest was renewed in the Galilean Collection from this point of view, and the need increasingly felt to continue the editorial task undertaken a century before by Antonio Favaro. It is certainly no coincidence that the project has been continued with one of the most important and least known sections of the Galilean Collection, the Correspondence of Galileo's Followers.

As Eugenio Garin stressed more than twenty years ago, 'without letters, the culture of certain centuries would be incomprehensible. They are interconnected collections of essays; they are diaries to be put back in order as a whole; they are chapters of collective books waiting to be rearranged; a variety of critical comments and points of view, each to be put in its place.'[30] In the age of the Scientific Revolution, communication by letter became the symbol of the construction and assertion of a new culture which recognised neither national boundaries nor religious divisions.

The first two volumes edited by Paolo Galluzzi and Maurizio Torrini, which bring together the letters preserved in the Galilean Collection covering the period from 1642 to 1656,[31] are an irreplaceable source for anyone who wishes to deepen their knowledge of events and figures connected with the new science of the seventeenth century: Torricelli, Cavalieri, Niceron, Carcavy, as also Boulliau, Holstein, Viviani, Ricci and Borelli, are but a few of the leading figures in this dense and remarkable interchange.[32]

After a long interruption, the work on the Followers' Correspondence, totalling in all ten thousand letters, has now finally been recommenced. This editorial task is part of a more extensive project for the electronic cataloguing and recording on

28 On the political culture of Fascism regarding science, see P. Galluzzi, 'La storia della scienza nell'E42', in T. Gregory and A. Tartaro (eds.), *Utopia e scenario del Regime. Ideologia e programma per l''Olimpiade delle civiltà* (Venezia, 1987), pp. 53–69; G. Baroncelli-M. Bucciantini, 'Per una storia delle istituzioni storico-scientifiche in Italia. L'Istituto e Museo di storia della scienza di Firenze', *Nuncius*, 5 (1990), 5–52; P. Galluzzi, 'Le istituzioni storico-scientifiche', in P.L. Ballini (ed.), *La Toscana nel secondo dopoguerra* (Milan, 1991), pp. 879–97.

29 See P. Pagnini (ed.), *Opere dei Discepoli di Galileo. L'Accademia del Cimento*, part 1 (Florence, 1942). Pagnini also prepared the critical edition of the experimental diaries of the Cimento Academy which, however, was never published.

30 E. Garin, 'Preface', in P. Galluzzi and M. Torrini (eds.), *Le Opere dei Discepoli di Galileo* (n. 5), i, xiii. On the importance of the correspondence in the age of Scientific Revolution, see P. Dibon, 'Communication épistolaire et mouvement des idèes au XVIIème siécle', in G. Canziani and G. Paganini (eds.), *Le edizioni dei testi filosofici e scientifici del '500 e del '600* (Milan, 1986), pp. 73–88.

31 See Galluzzi and Torrini, *Opere dei Discepoli di Galileo* (n. 5).

32 On the Followers' correspondence see M. Torrini, 'Il carteggio dei discepoli di Galileo', in G. Canziani and G. Paganini (eds.), *Le edizioni dei testi filosofici e scientifici del '500 e del '600* (Milan, 1986), pp. 167–81; U. Baldini, 'Verso una definizione storica della "filosofia" del galileismo: gli epistolari come strumento interpretativo', *Rivista di storia della filosofia*, 42 (1987), 213–35.

CD-Rom of a significant part of the autograph papers belonging to Galileo and the Accademia del Cimento.

The project's main goal is to create a bank of information and high-resolution images thanks to scanning procedures which will allow all scholars selective access to the documents in question, according to their interest. To this end, all necessary tools will be provided (indexes, bibliographical and lexicographical information), which will make it possible to move around easily while confident of reaching the intended research goals.[33]

As for the correspondence, the project proposes to make available a substantial part of the letters of the Cimento years (1657–67: files 275–282, about 2500 pages) both in electronic format and in print, while, in a second phase, it will publish the entire epistolary collection up to 1703, the year of Viviani's death.[34] This project, put into effect by the Florence Central National Library with the assistance of the Institute and Museum of the History of Science, is especially significant, not only because of the importance of the Collection itself, but also because, once completed, it will provide a useful point of comparison with other similar initiatives of equal bulk and complexity which are being carried out.

I am convinced that the study of the Galilean Collection, and our understanding of his manuscripts and of his correspondence, is still far from complete, especially now that research perspectives have multiplied and there is no longer a single privileged research centre but many co-existing historiographical routes and directions. The new interpretations of Galileo – as courtier, rhetorician or epistemologist – which have sprung up in the last decade, alongside the more traditional ones of Galileo the heretic, or Galileo the mathematician and philosopher, will really be put to the test in the Florentine Collection. Like the study of the birth of the first scientific academies, which is becoming so important and successful, the investigation of the relationship between science and political power, science and Inquisition, science and censorship, will certainly find new elements for discussion and food for thought in the Galilean Collection.

There is, however, one area in which the scholar of Galileo and the Scientific Revolution will perhaps be disappointed. I am referring to the Galileo affair, which remains of one of the most talked about and controversial questions in Galilean research. As far as this matter is concerned, one must go beyond the Galilean Collection and look in other directions.

The so-called reinstatement of Galileo by Pope John Paul II on 31 October 1992 has been widely discussed. It is not appropriate here to stress the clearly symbolic

33 As well as this project, another important work, also concerning the Galilean Collection, is reaching its completion. I refer to the electronic edition of MS Gal. 72, which contains one of the most widely discussed manuscript sources of Galilean historiography, namely Galileo's studies of motion. The project, arising from a collaboration between the Institute and Museum of History of Science of Florence, the Max Planck Institute of Berlin and the National Central Library of Florence, is edited by Jurgen Renn and Peter Damerow.

34 On the history of the Accademia del Cimento, see P. Galluzzi, 'L'Accademia del Cimento: "gusti" del Principe, filosofia e ideologia dell'esperimento', *Quaderni storici*, 48 (1981), 788–844; id., 'The Renaissance Academies: a commentary on Sessions I and II', in T. Frängsmyr (ed.), *Solomon's House Revisited* (Canton, Ma., 1990), pp. 303–21, on pp. 310–14.

significance of the Catholic Church's adoption of such a stance, when it is increasingly concerned about the direction and the research programmes which characterise contemporary science (especially in the field of the life sciences and genetic research). However, the matter certainly deserves attention. From this point of view, I believe that the act which would most truly signify real renewal and discontinuity with the past has yet to take place: that is, to allow historians, any historian who so wishes, the freedom to carry through their work by gaining access – the access granted in all the world's archives – to the manuscript collections preserved in the archive of the ex-Holy Office. Then, Galileo, *all* of Galileo, may once again be delivered up to historians for discussion and judgment; then, at the end of the second millennium, that archive, or at least a significant part of it, might at last be numbered among the archives of the Scientific Revolution.

3

Archive Refractions: Hartlib's Papers and the workings of an Intelligencer

Mark Greengrass

The papers of Samuel Hartlib, conserved in the University Library, Sheffield – some 25,000 manuscript pages and a further 1,000 pages or so of printed materials – are not a 'scientific archive' in the way that we might conventionally conceive of such things. Firstly, Samuel Hartlib was not a natural philosopher or a practising scientist, in whatever fashion we may define and apply these categories, anachronistically or not.

Contemporaries regarded him, and he projected himself, as a 'conduit-pipe' of knowledge, the 'hub of the axletree of knowledge' and as the 'great intelligencer of Europe'.[1] The first half of the seventeenth century was the great age of the 'intelligencer' in the advanced civil societies of western Europe. These individuals took advantage of the strong traditions of scribal publication. They worked outwards from the information networks of great merchant houses or ducal princes. They functioned in a world alongside newspapers, newsletter versions of which they sometimes prepared. They were not constrained by the existence of scientific societies who would, in due course, be able to channel scientific communication; instead they provided pathways by which it could be sustained at a relatively limited level across the boundaries of religious or political divisions.[2] Hartlib and others recognised the distinctiveness of the role of intelligencers through their attempts to establish a more multifaceted version of the Bureau d'Adresse which Théophraste Renaudot had set up in Paris and various provincial French cities.[3] Although Hartlib never succeeded in persuading the English

1 '. . . I finde my selfe obliged to becom[e] a conduit pipe . . . towards the Publick': Hartlib, 'Preface to the Reader', in *A Discours of Husbandrie used in Brabant and Flanders* (London, 1650). For other appreciations of Hartlib's role as an intelligencer, see G.H. Turnbull, *Samuel Hartlib. A Sketch of his Life and his Relations to J.A. Comenius* (London, 1920).

2 There is no satisfactory study of the role of 'intelligencers' across Europe; but of their importance to the increasingly pluralist flows of information in this period there is little doubt.

3 For the Office of Address in its various guises, see Charles Webster, *The Great Instauration* (London, 1975), pp. 67–77. On Renaudot, see H.M. Solomon, *Public Welfare, Science and Propaganda in Seventeenth-century France* (Princeton, 1972); additional clarification on the role of the Renaudot model is provided in Mark Greengrass, 'Samuel Hartlib: intelligenceur européen', in *Diffusion du savoir et affrontement des idées, 1600–1770* (Montbrison, 1993), pp. 213–34, esp. pp. 216–17.

Parliament to grant his Office of Address official recognition, still less regular public funding, Hartlib operated some components of its activities unofficially and individually.[4] In this respect, his archive is the fragmentary remains of that effort.

Secondly, Hartlib's papers do not constitute a scientific archive because his aspirations were much larger than merely a diffusion of knowledge about natural philosophy. He sought to overcome any narrow compartmentalisations of knowledge. His scope was all knowledge, physical and metaphysical. The Christian's duty was to restore the betrayed wisdom 'which Adam did bring into the world through lust'.[5] Communication of ideas was a godly discipline, by means of which every individual was required to contribute his talent to the good of all; 'every one as he hath received any gifft from God (which is a Character of his vertue and glorie) should communicat the same unto others'.[6] The 'duties of communication' were laid upon us all by Scripture. It was one of the ways by which knowledge could be turned to godly and utilitarian benefit for mankind in a 'Great Instauration'. So, when Hartlib and his collaborators, John Dury and Jan Amos Komenský, signed their famous 'pact' on 3 March 1642, they agreed to work together, even if separated by great distances, 'to the glory of God and utility of the public'. Such goals were to be met by pursuing religious peace in Europe, by educating Christian youth to the pathways of 'true wisdom', as well as by a reformation of all human knowledge away from the 'vanities of the world' and towards the 'verities of true happiness'. This is an archive in which its scientific concerns are intermingled with substantial educative, religious, occasionally millenarian or utopian refractions. Given the 'multiplex' nature of correspondence of this period, which often dealt with matters of natural philosophy alongside contemporary politics, or religious speculation or family matters, it is fatally easy, but dangerously flawed, to 'abstract' the scientific materials from a collection because of our interest in them, and thus ignore the delicate contexts in which they were being placed, or remove the contemporary relevance which the correspondent saw them as having at the time.[7]

The mention of John Dury and Jan Amos Komenský takes us to the third reason why the Hartlib archive is not a scientific archive. Every archive, but particularly one of correspondence and scribal communication, actively creates its own scribal shadow elsewhere whilst it, in turn, passively refracts a penumbra of archives

4 Hartlib was granted a pension of £100 per annum by the Long Parliament; this put him amongst the best remunerated of public servants of his generation, even though it required sustained diligence on Hartlib's part to secure its payment.

5 Hartlib Papers [hereafter HP] 47/10/4A ('A Further Discovery of the Office of Address', n.d., c. 1648).

6 Ibid., 47/10/17B.

7 An example of the (unintentional) harm done to a letter collection by selective abstraction is to be found in the recent edition of the remarkable correspondence of the factor and book-purchaser of Duke August the Younger of Braunschweig-Lüneberg, founder of the famous library at Wolfenbüttel. The letters of Philipp Hainhofer contain invaluable material for reconstructing the information networks of Europe in the time of Samuel Hartlib, but this is not readily apparent from Ronald Gobiet (ed.), *Der Briefwechsel zwischen Philipp Hainhofer und Herzog August. d. J. von Braunschweig-Lüneberg* (Munich, 1984), where it is only references to *objets d'art* and books which are abstracted from the letters to form the edition.

elsewhere. Archives do not exist in a vacuum, the faithful reflecters of the careers and concerns of their creators. They are also an amalgam of scribal imprints left by others. In the case of the Hartlib archive, it contains the long shadow of the treatises, letters and memoranda of his collaborators, most notably those with whom he signed the secret pact in 1642, Comenius and Dury, who together have been described as the 'the Philosophers of the Puritan Revolution'.[8] In their various ways, the disposition of the archival remnants in Sheffield superficially, but only partially, reflects the differing life-histories and aspirations of each of the three individuals. Comenius was an exile from his native Czech lands in the 1620s whose own personal library and papers suffered two consuming fires during his lifetime, the first during his hasty escape from Moravia in the years after the Battle of the White Mountain, and the second during the disastrous siege of Leszno in 1656.[9] After his early exile, he spent the remainder of his long life in a kind of eremitic existence, part invited guest, part wandering scholar, part visiting educational reformer and part figurehead for the other exiles from Czech lands, the majority of whom had taken up residence in southern Poland. What survives for Comenius, with the exception of his printed works, is therefore very substantially the passive 'shadow' of his own papers – the scribal imprint which it left in the archives of others. This imprint reflects both what they chose to keep as well as what he had chosen to send them. We can see this very clearly in the patterns of Comenius' surviving active correspondence, some 430 letters and fragments addressed to others in his name. These have been painstakingly calendared from across Europe by the team in Prague charged with editing his papers.[10] (See Figure 1) The result is a very variable survival rate, heavily dependant on chance. In some instances (such as his correspondence with Dury) we may detect the gaps. If this picture broadly reflects an enlarging pattern of Comenius' contacts throughout Europe, this was doubtless the result, at least in part, of the extraordinary success of his famous language textbook, the *Janua Linguarum*. Yet we can also detect in it the various networks in his communication – centring around Gdansk, the Czech Brethren in Poland, his long-standing patron Louis de Geer and his factor, and finally Samuel Hartlib and his associates. Hartlib had collected money for Comenius, and published (without Comenius' authorisation) the early treatise which Comenius had sent him, sketching out how 'pansophia' might be realised by a reformation of learning. In due course, he had publicised Comenius plans for this reformation

8 H.R. Trevor Roper [Lord Dacre], 'Three Foreigners: the Philosophers of the Puritan Revolution', in *Religion, the Reformation and Social Change* 2nd ed. (London, 1972), pp. 237–93.

9 For the details of Comenius' life, and (in particular) the survival of his papers, the most comprehensive study remains Milada Blekastad, *Comenius. Versuch eines Umrisses von Leben, Werk und Schicksal des Jan Amos Komenský* (Oslo and Prague, 1969).

10 These figures have been compiled from a communication by Vladimír Urbanek to the Czech Academy of Sciences Workshop on Intellectual Communication in the Seventeenth Century, held in September 1995, and due to be published in the forthcoming *Acta Comeniana* (1997). The figures include as a separate item those official letters sent by Comenius in his position as a senior of the Czech Brethren. I am very grateful for Dr Urbanek's permission to utilise his findings in advance of publication in this chapter.

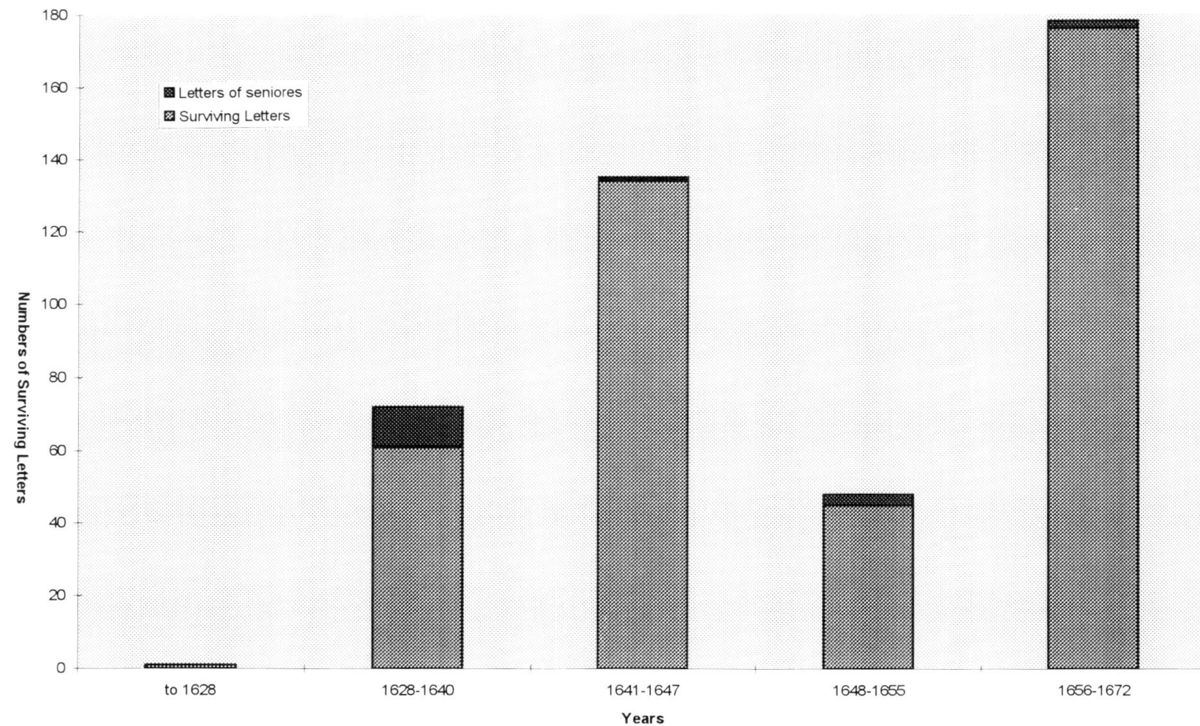

Figure 1. Surviving letters and fragments of correspondence originating with Comenius

of learning by sending out copies and inviting comment on them. Even now, the Hartlib archive contains more unpublished Comenius material than exists in Prague, and particularly from the 1630s and 40s when he was developing his distinctive philosophies of language and educative endeavour. Almost a quarter of his surviving correspondence is to be found in Sheffield. Although we may readily infer what it was in Comenius' writings which evoked such an enthusiastic response in Hartlib, the full significance of this material (for example, the substantial correspondence between Comenius and Cyprian Kinner which found its way into Hartlib's hands) is only becoming apparent as the remaining corpus of his correspondence is recovered from its shadow existence in the libraries of Europe.

There has been no counterpart to the effort of Comeniologists for the surviving papers of John Dury. Dury was the seventeenth century's answer to Terry Waite, the 'peacemaker without partiality', whose efforts were, as he saw it and as later historians have tended to agree, endlessly frustrated by quarrelsome, sectarian-minded continental clerics.[11] As with Comenius, there is no such thing as a 'Dury archive'. His substantial remaining papers in Sheffield come the nearest to constituting it in the sense that Hartlib's house (initially in Axe Yard, Charing Cross) was where he left trunkloads of papers when he went on his continental missions. Hartlib acted as his mail-box in London and also raised money for the exiled protestant groups with whom Dury had established contact, and whom he hoped to be able to deliver assistance to in the name of a common cause, or 'correspondency'. He organised for several of his treatises to be published. It was Dury who seems first to have encouraged Hartlib to regard the collection of scientific intelligence as of fundamental significance to the Great Instauration.[12] His surviving letters from the 1630s suggest that he was as good as his word when he offered to act as an unofficial collector of such intelligence since they frequently include a bewildering mixture of ecclesiastical politics, interspersed with casually acquired scientific comment or description. It was also John Dury who wrote the first prospectus for an Office of Address which Hartlib published after the first English Civil War in 1647.

Yet putting the Dury materials in the Sheffield collection into the context of other surviving, unpublished Dureana must be currently regarded as beyond us. We do not even possess as yet a definitive list of the various editions of the printed publications to be ascribed to him. Having begun to establish a calendar of unpublished Dureana, it currently runs to over 500 items and it is far from complete. It includes references to materials which I know to exist in Zurich, Oxford, Marburg, Geneva, Durham and elsewhere, but which I have not yet seen

11 See J.M. Batten, *John Dury, Advocate of Christian Reunion* (Chicago, 1944); and, for a recent revised assessment, Anthony Milton, ' "The Unchanged Peacemaker"? John Dury and the politics of irenicism in England, 1628–1643', in Mark Greengrass, Michael Leslie and Timothy Raylor (eds.), *Samuel Hartlib and Universal Reformation* (Cambridge, 1994), pp. 95–117.

12 J.M. Batten, op. cit., p. 28, citing J. Kvacala, *Korrespondence Jana Amosa Komenského* (Prague, 1902), pp. 6–7.

or properly verified. So the degree to which this substantial fraction of Hartlib's papers can properly be interpreted at the moment is hampered by the failure to recover the 'shadow' archive to which it relates. The effect of having only a 'partial' archive like this is to encourage a quarrying of material for particular purposes without regard to the overall picture. Thus, there has been a tendency to concentrate on Dury's connections with the Jewish communities in Europe, and on the millenarianism in his thought in the 1650s,[13] while relegating his efforts at 'scriptural analysis' along the 'technological lines' which he regarded as replicating the principles of the mathematical sciences in theology, and ignoring his concern for a 'right method' in all human enquiry. There is also a tendency not to notice the degree to which Hartlib was more than just his London agent, sometimes going well beyond the suggested or required brief which Dury had proposed for him in what he published or distributed under his name.[14]

So far we have been examining the shadows from elsewhere within the Hartlib collection. What of Hartlib's own papers? That the 72 bundles of papers now extant only contain a fraction of the original, and much more extensive, archive is reasonably evident. Within months of his death, Hartlib lamented to his friend John Worthington on the 'distraction or embezzlement' of many of his books and manuscripts as a result of entrusting them to an unnamed friend for safekeeping.[15] More were apparently lost as the result of a fire in his house.[16] It may well be that this fire took place in, or around, 1657 – in which case it may explain (at least in part), the pattern of survival of Hartlib's passive correspondence in the collection revealed in the enclosed graph (Figure 2). Not too much reliance, however, should be placed on these figures since it was in the nature of correspondency networks that they developed their own internal dynamics which themselves explain the velocity of intelligence traffic at any particular moment. In his declining years, Hartlib moved to stay with his son, also called Samuel, in Angel Court, Westminster, a near-neighbour to another diarist – Samuel Pepys. Whilst he was there, and in poor health, his friend Samuel Wartensky was alarmed to find that his friend's precious papers were 'a prey to plunder by all'.[17] The process of selective degradation clearly continued after his death in March 1662, when the remaining archive was purchased by William Viscount Brereton and transported to his country seat in Cheshire. It was there that John Worthington had providentially come upon them and put them in order:

13 See, for instance, R.H. Popkin, 'The end of the career of a great seventeenth-century millenarian: John Dury', *Pietismus und Neuzeit*, 14 (1988), 203–20; id., 'Hartlib, Dury and the Jews', in Greengrass, Leslie and Raylor, *Hartlib and Universal Reformation* (n. 11), pp. 118–36

14 One example of the extensive, and 'creative' expansion of Dury's ideas for publication is to be found by comparing the original letter to Hartlib (HP 47/14/1–8B) and the subsequent treatise on *The Reformed Librarie-Keeper* (London, 1650), published in Dury's name, and on which it was based.

15 John Worthington, *Diary and Correspondence*, ed. J. Crossley, 3 vols., Chetham Society (Manchester, 1847–86), ii, 67.

16 Ibid., ii, 107 (6 February 1662). See also Turnbull, *Samuel Hartlib* (n. 1), p. 72.

17 'omnium exposita rapinæ' – HP 32/3/40A – Wartensky to Hartlib, 23 July 1661.

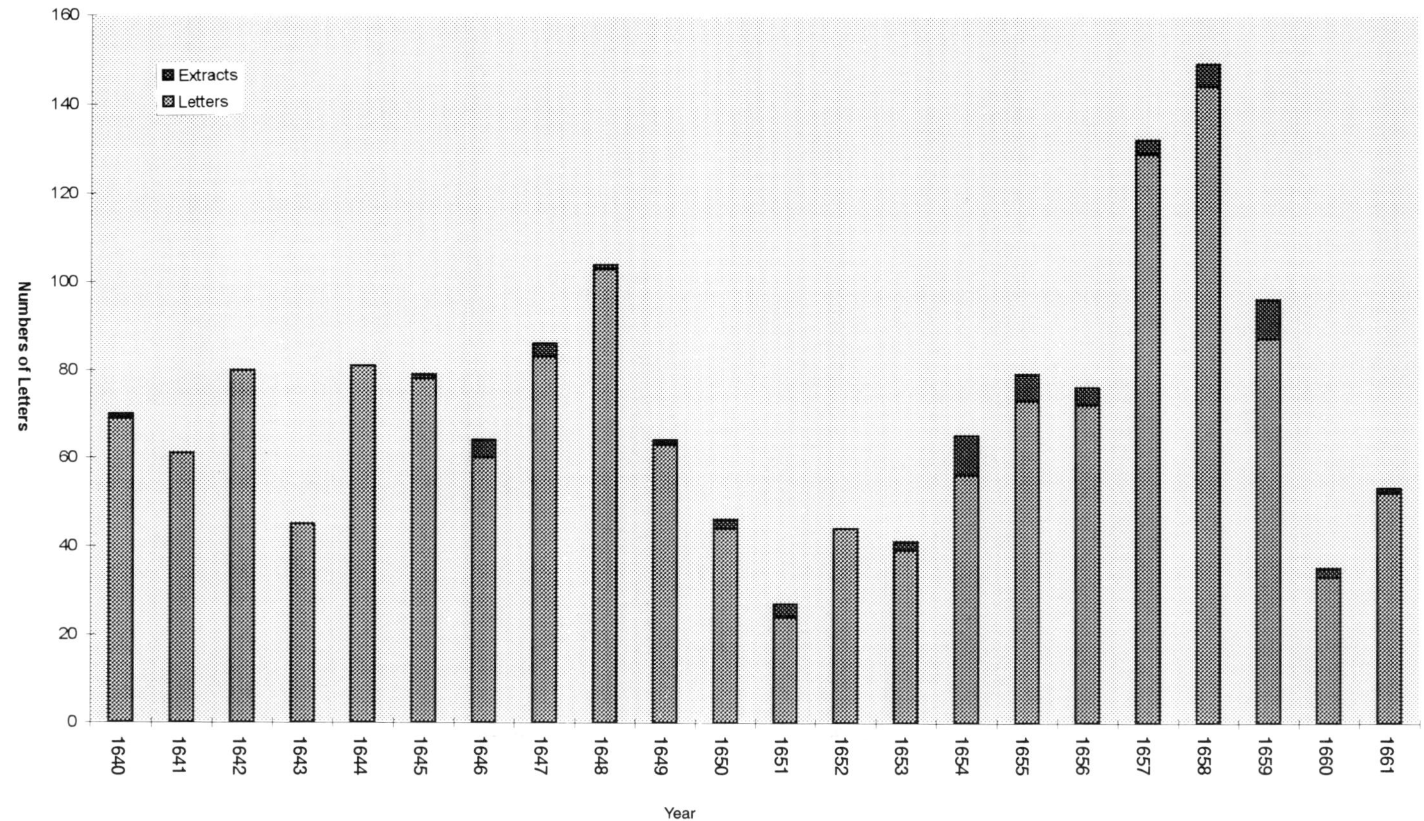

Figure 2. Surviving letters to Samuel Hartlib, 1640–60

> And though my latter years have had something of the pilgrim state, yet I have had cause fo acknowledge some merciful designs of providence therein. At my late being in Cheshire I met with two trunks full of Mr. Hartlib's papers, which my Lord Brereton purchased. I thought they had been put in order, but finding it otherwise, I took them out, bestrewed a great chamber with them, [*and*] put them into order in several bundles . . .'[18]

In the process, some papers were almost certainly abstracted by Worthington. His own letters were inevitably removed (to be copied later as part of Worthington's own manuscript collection). Those between Seth Ward and Hartlib probably disappeared at the same time. In a letter to Worthington, shortly after his cataloguing of them, Ward remarked:[19]

> I am very glad that those papers of Mr Hartlib's are preserved and that they are fallen into your hands who are able & disposed to make the best of them; I was not unacquainted with that good man, who by his great and unwearied Zeale for learning, and by his Correspondence with persons eminent, in the severall wayes of it became serviceable to the generall promotion of it; and whatever his workes (w*hi*ch were very laudable) certainly it cannot be but his papers must be considerable. I meane those papers which proceeded from the Authors whome y*ou*r letter mentions, and not those letters of mine owne w*hi*ch concerned either Hevelius or Mercator, w*hi*ch although I have forgotten, yet so much I am sure of, that they were carelessly & perfunctorily written (or els indeed they had not been mine) so that it wil be to my advantage to suppress them. However S*i*r! I leave them wholely to y*ou*r disposall, either to bring them to me when I may have the happiness to se you, or to burn them or leave them among the rest that is to say I have no considerable regard for any interest of mine in them.

It is possible that others had ransacked the collection and removed materials which might have presented their involvement in the affairs of the Commonwealth and Protectorate in not quite the uncommitted light that they hoped to put abroad *post facto*. Even John Milton, or so it has been suggested, may have taken steps to ensure that materials in the collection relating to himself were abstracted. Some of Hartlib's papers appear to have passed through the hands of Hans Sloane into the British Library collection, and presumably from around this period. One whole bundle was dispersed, probably around this time, and purchased for the Osborn Collection of the Beinecke Library at Yale University in 1957.[20] And, although the collection disappears from view in the later 1660s, only to re-emerge with its

18 Worthington, *Diary and Correspondence* (n. 15), ii, 230.

19 BL Add. MS 6271, fol. 8.

20 One of the Petty letters in the Yale bundle is endorsed in the same hand that endorses other bundles in the Sheffield Hartlib Papers – 'Petty Mixt Letters & Papers of no great value [th]at I know of'. The hand may well be that of Worthington himself but it has proved impossible to authenticate it. The only specimens of Worthington's autograph which I have identified are from the audit books of Jesus College, Cambridge, where he was master, and his supplications for his degrees at the University of Cambridge. I am grateful for the assistance of Dr Frances Willmoth, archivist of Jesus College, and Dr Mark Nicholls, university archivist, in attempting to resolve this question.

discovery in 1933 in a London solicitor's office by the then professor of education in Sheffield, George Turnbull, this did not completely prevent further slight losses and at least one piece from the collection has turned up in a sale catalogue over the past fifty years.[21]

The inevitable winowings of time and fortune upon an archive are such that we come to expect and it should not automatically be assumed that our interpretation of the *surviving* material is radically affected by this albeit selective tampering. And, since Hartlib's activities attracted mild disfavour, quiet ridicule and then complete obscurity after his death, we do not have layers of Enlightenment patina to remove from the portrait of his activities. Perhaps it was still the case that Thomas Shadwell's *The Virtuoso could include the character 'Sir Samuel Hearty' as an intended caricature of Hartlib in 1676;*[22] if so, it was not noticed by contemporaries – or by modern commentators. He was remembered, if at all, only as the 'friend of John Milton', since it was to Hartlib that Milton had dedicated his tractate *On Education* in 1644. The rediscovery of Hartlib's archive, and with it, that of his activities, influence and significance, is a uniquely twentieth-century phenomenon. In that respect, the problems of refraction created by the survival of this archive are less significant than those relating to its structure.

Since Worthington reorganised the remaining papers, we clearly do not have an archive which reflects the structure which it had when Hartlib was in charge of it; and for someone who was habitually methodical in the collection and dissemination of information, it is surprising how little we know about how he organised his papers. The surviving evidence comes mainly in the form of bundle wrappers with brief annotations as to the contents of the bundle. The majority of these are in two hands, both of which are in manuscript hands which are near-contemporary to Hartlib, and one of which is likely to have been that of Worthington himself.[23] But there are some bundle wrappers in Hartlib's autograph and it seems likely from these wrappers that Hartlib organised his incoming correspondence by separate bundle, alphabetically according to correspondent, sometimes numbering the sequence of letters as they arrived.[24] Hartlib was a considerable scribal copyist and had a substantial collection of manuscript books which he lent out to individuals around his network on occasion.[25] These we know were numbered, presumably so that he could keep track of them. Beyond these, somewhat casual and physical signs of the organisation of his archive, there is the dense evidence from his own unedited diary or *Ephemerides* which provides considerable evidence for the mental scaffolding which he erected around his information-gathering network.

21 Information kindly provided by Dr Charles Webster.

22 Thomas Shadwell, *The Virtuoso*, ed. M.H. Nicolson and David Rodes (London, 1966), p. 8 and passim.

23 These hands are delineated as 'y' and 'z' in *The Hartlib Papers: a Complete Text and Image Database of the Papers of Samuel Hartlib (c. 1600–1662)* (2 CD-Rom discs and User Guide, Ann Arbor, 1995). E.g. 53/43/1A 'Project & Proposals', or 58/1A 'Proposals for sundry Bubbles from 1633 to 1660', both in hand 'z'; for an example of hand 'y', see 41/1/1A ('Letters from Mr Austen about planting').

24 E.g., HP 56/1/1A.

25 Referred to more extensively in Mark Greengrass, 'Samuel Hartlib and Scribal Publication', article forthcoming in *Acta Comeniana* (1997).

For Hartlib, as for Comenius, the keeping of a private diary was a commendable intellectual and spiritual endeavour. In the *Linguarum Methodus Novissima*, Comenius had advised all those who wanted to be 'rationally engaged in life' ('rationabiliter in vita versari') to keep a 'vitæ Ephemerides', a book in which the events of each day would be recorded.[26] He added that such a diary should include three things wherever possible. There should be a note of whatever one had read which had made an impression; of anything which had been overhead in conversation which was of interest, and, finally, whatever individual ideas, thoughts or reflections came to one which were worth recording. Such a diary would, when read and re-read, lead one closer towards God and His world of Creation and the amendment of one's life. Hartlib was similarly impressed with the importance of keeping a daily diary. In this respect he probably reflected contemporary English Puritan practice which regarded the keeping of a record of God's providence as a godly way to achieve peace of mind.[27] For Hartlib, however, a diary had a *scientific* and a *spiritual* – eventually a pansophic – purpose As he wrote at one point in it: 'The great means to come insensibly to a vniversal knowledge and experience is to keepe diaries exactly of all whatever wee heare or see by way of converse out of Books'.[28] Politicians, physicians, astronomers, even children, were encouraged to keep diaries as a discipline 'to make all men easily Pansophicall'.

Hartlib's own diary runs to over a third of a million words in its surviving manuscripts. It was written entirely by Hartlib and he kept to a regular format throughout. Entries were transcribed into a series of unsewn manuscript books in which a regular running margin was used to insert an index to the contents and the persons mentioned. There is every indication that these books were copied up from more personal day-to-day diaries or from correspondence collected together over a period of several weeks or even months. This was an essential part of Hartlib's intelligencing activities, a 'tool of the trade'. He clearly regarded it as valuable and there are internal signs that he consulted it from time to time to find references to books or addresses which it contained. Since he never provided the diary with a general index to its contents, however, we must imagine that it was in the act of writing it, the sorting and epitomising process which it represented, that its worth essentially resided for him. The surviving manuscript begins in 1634, but there is no sense that this was when it had been begun; and it is more than likely that he commenced the diary in, or around, 1631, when Dury started his irenic continental sojourns. Thereafter it continued until 1660, but there are gaps in the surviving manuscript record from 1636 to 1638, and again from 1644 to 1647. There are also some missing quires within those years for which the manuscript is extant. Overall, however, it constitutes the best evidence we have for how Hartlib saw his own archive.

26 'librum scilicet, in qvo singulis diebus acta sua consignet' – HP 35/5/154B.

27 See Paul Delany, *British Autobiography in the Seventeenth Century* (London, 1969), esp. pp. 63–6; for the theological justification for diary-keeping, see Dean Ebner, *Autobiography in Seventeenth-Century England. Theology and the Self* (The Hague/Paris, 1971).

28 HP 31/2/10A–B (Ephemerides, 1648).

Firstly, the diary contains a good deal of information about the individuals who made up his 'network'. Each entry is followed by the name of the correspondent or source which provided the data recorded. From these names, it is tempting to reconstruct, albeit with substantial reservations, moving profiles of Hartlib's collaborators and acquaintances and compare it with the similar moving profile of correspondents provided by the surviving letter collection. It is a complex picture which ideally should be replicated by clock-dial organigrams on a year-by-year basis, the more frequent correspondents being closer to the centre of the dial. The pattern is broadly the one which is to be expected of an 'inner wheel' of regular correspondents and a penumbra of more intermittent communicators. Social class was not a determinant of the accuracy of information; corroboration (experimental veracity, or apparently proven medical worth) was. Individuals are not classified by religious belief or political persuasion. Barbers, servants, maids, carriers and unskilled artisans were as likely to have profound, if unpolished, secrets of nature; and a true intelligencer was duty bound not to ignore information coming from all sources, including the least likely.

Hartlib treated books and manuscripts as objects not to be pondered but plundered. Books were to be methodised, epitomised, indexed, gutted of their information. He did not want to own them, but he wanted to record information from them of possible use to others. He regarded all worthwhile knowledge as hidden, subterranean and precious. The adjectives he applies to the books of which he approves in the *Ephemerides* are telling. They are 'very rare', 'exquisite', 'precious'. As Stephen Clucas has demonstrated, Hartlib had a peculiar fascination for the hidden knowledge which might be secreted in manuscript collection, a reflection of his own involvement in scribal publication as a means of intellectual communication.[29] He collected scribal imprints. He was fascinated by the process by which books could be indexed and summarised and libraries catalogued so that their information could be more generally available.

Hartlib's own structural divisions of knowledge are indicated by the marginal subject-matter citations of the *Ephemerides*. These are worth a closer analysis since they reveal the evolution of Hartlib's interests over the period from the 1630s to the late 1650s. Here was an Encyclopaedia in the making: a glimpse inside the mental filing cabinet as the materials were being organically assembled. The categories varied over time, with 'Libri selecti Theologici', 'Practica divinitas', 'Pansophica didactica' and 'Logica' being amongst the most frequently cited categories in the 1630s. By the 1650s, these had become, to a degree, supplanted by 'Oeconomica', 'Oeconomica Physica', 'Experiencia Mechanica', 'Chymica', 'Opera Parliamenta' and 'Didactica divitiarum'. Above all, the diary reveals that Hartlib had an awareness of method itself. His quest for a way of cataloguing his information into the three books of 'nature' the 'revealed word of God' and 'mental ratiocination' led him to collect different views on just how much the 'praecognita'

[29] Stephen Clucas, 'Samuel Hartlib's Ephemerides, 1635–59, and the pursuit of scientific and philosophical manuscripts: the religious ethos of an intelligencer', *The Seventeenth Century*, 6 (1991), 33–55.

of human knowledge should be used to shape the ‘experientia’ of the physical world as it appeared to us. In this respect, the earliest passages of the diary, those from 1634 and 1635, are amongst the most complex, most difficult to interpret, and the most fascinating.

Knowledge of how Hartlib’s mental archive worked is significant because it assists us to read his real archive in a different light. Once one realises at Hartlib distinguished between knowledge which had been transferred by the written word (which he sometimes calls ‘Historica’ – the accumulated wisdom from the past) and knowledge which has been acquired experientially (‘Experientia’; ‘Experimenta’; ‘Scientia’), it makes sense of the way in which he took his notes. Hartlib has often been accused of being ‘naive’ and ‘uncritical’ in his attitude to the intelligence he received across his informal information network. Yet such a misconception stems from a misconception about Hartlib’s metaphysic of knowledge. He distinguished, like Comenius, very sharply between the received wisdom of the written and the real world on the one hand, and its critical appraisal, the mental appreciation of such wisdom, on the other. So it is under headings such as ‘Pensa’ or ‘Critica librorum’ that Hartlib records his (or other people’s) views of particular works or experiments, or records the judgments of others whom he respects. Far from being uncritical, Hartlib could be severe, especially on works which proclaimed more than they delivered. Of Descartes, he could say: ‘Hee also is too much bragging. For hee promises more in his general dicourse [the *Discourse on Method*] than he dose [*leg*: does] performe.’[30] He was particularly incensed by any crudely mechanical or circus-like demonstration of the hand of God in nature. To treat God’s providence like a performing dog would be to encourage, particularly in the minds of children, a kind of idolatry. He was also temperamentally and intellectually opposed to an over-confident, over-arching, rash synthesis. God’s creation was infinitely more subtle and complex than we could possibly imagine. As human beings, we would only do damage to it, and to ourselves, by foolishly and vainly imagining that we could come swiftly and simply to a complete understanding of it. The particular emphasis in Hartlib’s circle upon ‘philosophia libera’ meant accepting a rather humble and modest pluralism of method and approach to the study of the physical world – as also in religious matters. In one remarkable passage of his diary, Hartlib compares the fortunate fructiferousness of such pluralism to the overflowing of a river: ‘It is not good to enslave ones-selfe to any kinde of Method or Meditations but to observe a certain generosity and liberty in all our Stud[ie]s. This will bee found to bee far more profitable. As the overflowings of rivers doe bring in a world of things with them which they never would have done, if they had always runne in their wonted channels . . .’.[31] Such beguilingly modest assertions of the benefits of a (relative) human ignorance would, in the conditions of intellectual change in mid-seventeenth-century Europe, have their undoubted uses. One begins to grasp the

30 HP 30/4/18B (Ephemerides, 1639).
31 HP 30/4/21A (Ephemerides, 1639).

beguiling nature of Hartlib's perception that it could only be be a corporative, public, shared, pluralist endeavour which would bring about a Great Instauration.

Whether his mental perception of knowledge was ever effectively transferred to the real archive is probably doubtful. He certainly does not seem to have made any systematic attempt to do so beyond the primitive filing systems already indicated. His mental construct of his archive did not necessarily refract upon the museum of paper in a busy intelligencer's repository. In just the same way, my beautifully constructed mental picture of my own papers is certainly not transferred in the more organic (possibly even 'chaotic') reality of the filing cabinets in my office. Although Hartlib undoubtedly appreciated the relevance of cross-referencing information, he did not undertake such cross-referencing himself. Although he would see the significance of regular updating of information, I doubt if he actually carried that through with respect to his own papers.

And what of the scribal shadow of Hartlib's archive left elsewhere? Theoretically, we should expect this to be very substantial; intelligencers generated outgoing paperwork in considerable quantity. After the first text and image edition of the Sheffield archive was completed in late 1994, I was given a grant for a year to search for this 'hidden half'. It was a frustrating and thankless task. Whilst Hartlib's autograph hand is very distinctive, the papers which he sent out from his office tended to be the work of his amanuenses – ten, possibly eleven, different unascribed hands (enumerated A–K in the edition). In many instances, it was a matter of looking for materials despatched by Hartlib but neither originating with him nor in his hand. The exercise brought a measure of success in the obvious places – the Hevelius correspondence in Paris, for example, revealed some letters to Hartlib which had not been found amongst his own papers. There were fragments in the Jungius archive in Hamburg, other oddments in more unlikely places (such as Chetham's Library in Manchester). Overall, however, the exercise has been fascinating and disappointing. There were all sorts of reasons why individuals should want to keep a letter from Comenius; it is not so clear why they might choose to keep one from Hartlib. The information which flowed across his network they absorbed and utilised; they did not need to keep the evidence. His papers remain for us as historians of seventeenth-century science like bright things found by a jackdaw, worth rifling through for hidden gems. The CD-ROM edition of his papers enables us to do more than rifle through them still more efficiently. It provides us with a means whereby we can reconstruct the mental archive which he had perhaps conceived, but whose reality had certainly eluded him.

4

Between Erudition and Science: The Archive and Correspondence Network of Ismaël Boulliau

Robert A. Hatch

> There is a line among the fragments of the Greek poet Archilochus which says: 'The fox knows many things, but the hedgehog knows one big thing'.
>
> Isaiah Berlin, *The Hedgehog and the Fox*

The 'Scientific Revolution' – unless reports have been greatly exaggerated – has recently undergone considerable down-sizing. Once the precocious child of a new discipline, the periodisation has long been heralded as the *locus classicus* of the History of Science, and as firstborn, it continues to reflect disciplinary as well as historiographic change. As the number and variety of practitioners has grown, as philosophers and sociologists have joined historians in asking new questions, the Scientific Revolution has undergone a simultaneous broadening of scope and narrowing of focus. Inevitably, perhaps, specialisation has also brought a growing sense of fragmentation, and arguably, monuments that once announced its birth now mark its demise. The elegant monoliths of Koyré, with their classic lines, once seemed adamantine, homogeneous, eternal; yet even the terrestrial visions of Butterfield – whose 'bridge' once challenged the gulf between the Two Cultures – have now fallen to local traffic and the vistas of 'small pond' culture. Perhaps what once tested the reaches of space and time – what once spanned 'European' and 'Western' as it passed over 'Renaissance' and 'Reformation' into 'Modern' – was nothing less than 'early modern science'.

But what summons such heroic visions? What, after all, do 'Archives of the Scientific Revolution' have to do with historiography and metaphor? The answer, of course, is that the Scientific Revolution has always stood on the shoulders of giants. Central to the vision were the great texts – the *De Revolutionibus*, the *Principia* – and of primary importance, the great standard editions – the *Opera omnia*, the *Oeuvres complètes* – all accessible, after generations of scholarly effort, in any good library.[1] Like learned colleagues, these monuments directed disserta-

1 Traditional monuments include Charles Adam and Paul Tannery (eds.), *Oeuvres de Descartes*, (13 vols., Paris, 1897–1913) and René Descartes, *Correspondance*, ed. Charles Adam and Gaston Milhaud (8 vols., Paris, 1936–63); Pierre de Fermat, *Oeuvres de Fermat*, ed. Paul Tannery, Charles Henry, and

tions and provided new avenues of research. But it was not without reason that the traditional heros of the Scientific Revolution – Copernicus, Kepler, Galileo, Descartes, Huygens, Newton – were so honoured and thus honoured first. Even a self-blinded scholar – desperately seeking shape in 'early modern science' – could not go untouched by such genius. There was, of course, more to the story, a broader context that cut across generation, class, temperament and traditional periodisations. Here we are reminded of the intelligence and industry of lesser lights, of a Mersenne, Hartlib, or Oldenburg. For the first time we see – even in Galileo's shadow – the outline of a working tradition,[2] and now joining the litany, an unlikely couple, Hobbes and Flamsteed.[3] Indeed, as new voices join the chorus, a certain dissonance – even shrillness – can be detected. But arguably it never was an ideal world, and for all that, harmony would never play in a real revolution. Here we need only recall voices from an earlier generation of humanists and gatekeepers, such honoured citizens of the Republic of Letters as a Chapelain or, for present purposes, a Peiresc.[4]

But if our vision of the Scientific Revolution has changed – if the great editions have continued to broaden and sharpen our image – can Archives contribute to a more satisfying 'Big Picture'? Archives, after all, are the traditional haunt of the biographer, and doubtless each bears a unique physiognomy. Nor are archives random collocations of paper; they have a history, they share a certain logic and family resemblance. The challenge is to include more archives, to draw together the various portraits and archival kingdoms into a larger picture – not a vision captured in cliché or resolved in abstraction, but a working map of intellectual commerce. In the simple narrative that follows, I trace the basic contours of Boulliau's correspondence network and the fate of his archive. I begin by situating him in the Republic of Letters.

Science, Learning and the Republic of Letters

When he was elected to the Royal Society in 1667, Ismaël Boulliau (1605–94) was over sixty years old. An early Copernican, Keplerian, and defender of Galileo, Boulliau's contribution was to an earlier age, to the halcyon days of the Mersenne

Cornélis de Waard (5 vols., Paris, 1891–1922); Galileo Galilei, *Opere*, ed. Antonio Favaro (20 vols., Florence, 1890–1939); Huygens, *Oeuvres complètes*; Johannes Kepler, *Gesammelte Werke*, ed. Walther van Dyck, Max Caspar, and Franz Hammer (Munich, 1937–), and *Joannis Kepleri astronomi opera omnia*, ed. Christian Frisch (8 vols., Frankfurt, 1858–71); *Newton Correspondence*, and Whiteside, *Mathematical Papers*.

2 *Le Opere dei discepoli di Galileo Galilei*, Ed. Nazionale, ed. Paolo Galluzzi and Maurizio Torrini (2 vols., Florence, 1975–84).

3 Thomas Hobbes, *Correspondence*, ed. Noel Malcolm (2 vols., Oxford, 1994); *The Correspondence of John Flamsteed, The First Astronomer Royal*, compiled and edited by E.G. Forbes, and (for M. Forbes) by L. Murdin and F. Willmoth (2 vols., Bristol and Philadelphia, 1995–97).

4 Editions include Nicolas-Claude Fabri de Peiresc, *Lettres de Peiresc*, ed. Philippe Tamizey de Larroque (7 vols., Paris, 1888–98), and *Les correspondants de Peiresc, Lettres inédites.* (2 vols., reimpression, Geneva, 1972); Jean Chapelain, *Lettres de Jean Chapelain*, ed. Philippe Tamizey de Larroque (2 vols., Paris, 1880–83).

Circle, the Cabinet Dupuy, and the first assemblies that foreshadowed state-sponsored science. An able astronomer, Boulliau published widely in science and mathematics as well as in history, classical studies, and theology. He was also active in the Republic of Letters. As Harcourt Brown suggested some sixty years ago, his 'lively mind reflects most of the movements, literary, social and philosophic, of his age'.[5] Of particular interest here, Boulliau was a prolific correspondent situated *entre deux âges*. Trafficking in information, news, and gossip – scientific, literary, and political – his contribution was to the generation between Mersenne and Oldenburg, to the passage from *érudit* to *homme de science*, from intelligencer to the scientific journal. Evidence of his activities is abundant. His letters survive in numbers that still stupefy.

The trick, of course, is to bring his manuscripts into meaningful order. Several decades ago the Annales historian, Robert Mandrou, suggested a simple strategy for comparing correspondence networks. The ingredients are simple enough. Mandrou traced the contours of the Peiresc network from existing lists of correspondents at Carpentras.[6] In an effort to establish a broader context for Boulliau's archive, what follows in outline is a simple comparison of his correspondence network with three others, specifically, the published letters of the N.-C. Fabri de Peiresc (1580–37), Marin Mersenne (1588–48), and Henry Oldenburg (1617?–77). For Boulliau, I have employed a calendar of the *Collection Boulliau*.[7] The result is a crude outline in *terra firma*. Plotted over space and time, the letters situate the epistolary efforts of four generations of intelligencers. Drawn together, the correspondence networks provide a glimpse of the Republic of Letters in transition.

N.-C. Fabri de Peiresc was the most prolific correspondent of his age. Situated in the south of France in the early decades of the century, Peiresc's career marks critical transitions in space, time, and intellectual interest. Though the exact size of his network is far from clear, at his death there were as many as 10,000–14,000 letters or register copies. A full reckoning of his extant and lost letters has not been undertaken. In 1973 Mandrou mapped Peiresc's correspondents – some 500 – using the Registers at Carpentras.[8] For purposes of comparison, I have plotted Peiresc's published letters, which shows some 3200 letters for the years 1598–1637 [Fig. 1]. Although over 100 cities are represented, nearly half involve Paris.

5 Harcourt Brown, *Scientific Organisations in Seventeenth Century France (1620–1680)* (Baltimore, 1934, reprinted New York, 1967), p. xvii. I wish to acknowledge the influence and personal support of the late Harcourt Brown.

6 Robert Mandrou, *Des humanistes aux hommes de science (XVIe et XVIIe siècles)* (Paris, 1973), 'Cartes' [pp. 248–9]; English translation as *From Humanism to Science 1480–1700* (Harmondsworth, 1978), 'Maps' [pp. 310–11]. Robert A. Hatch, 'The Collection Boulliau: A Quantitative and Comparative Study with the Correspondence of Mersenne and Oldenburg', History of Science Society Annual Meeting, Toronto, 1980 (unpublished). This paper was motivated in part by the claim that Boulliau 'rivaled Mersenne as a correspondent' (Carl Boyer, 'Boulliau', *Dictionary of Scientific Biography*, ed. C.C. Gillispie (16 vols., New York, 1970–80), ii, 348–9, on p. 349 col. 1).

7 The importance of systematic calendaring cannot be overstated; see Morris L. Radhoff, 'A Guide to Practical Calendaring', *American Archivist*, 11 (1948), 123–40, 203–22. The Boulliau calendar is modeled after 'Nijhoff model', employed by Hans Bots et al. in calendars for Rivet and Gronovius.

8 Mandrou illustrated the cities of correspondents rather than number of letters from cities. Cf. Mandrou, *Des humanistes* (n. 6), 'Cartes' [Appendix].

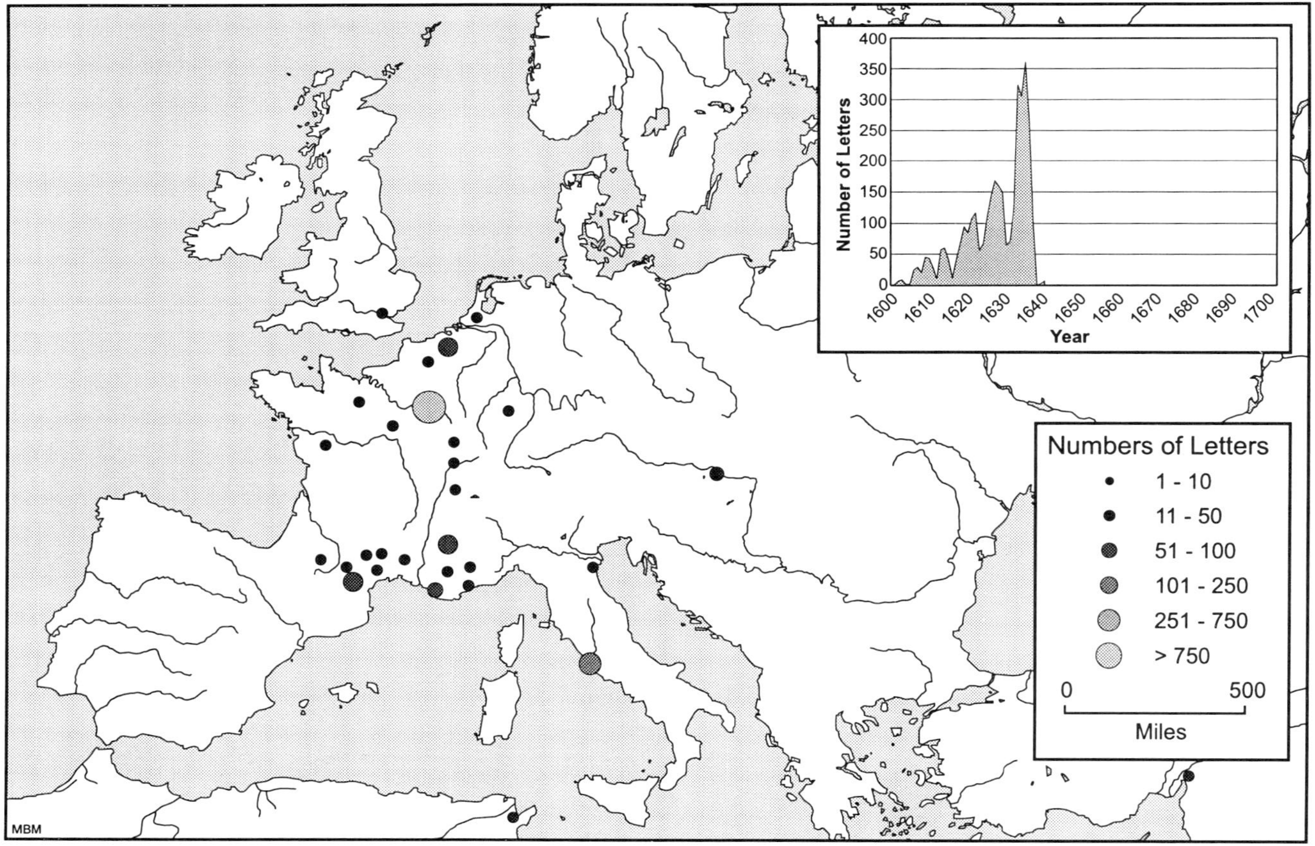

Figure 1. *The Published Correspondence of Peiresc*: Nicolas-Claude Fabri de Peiresc (1580–1637) left an estimated 10,000–14,000 letters at his death; perhaps half that number is extant. The present map and chart represent some 3,200 published letters for the years 1598–1637.

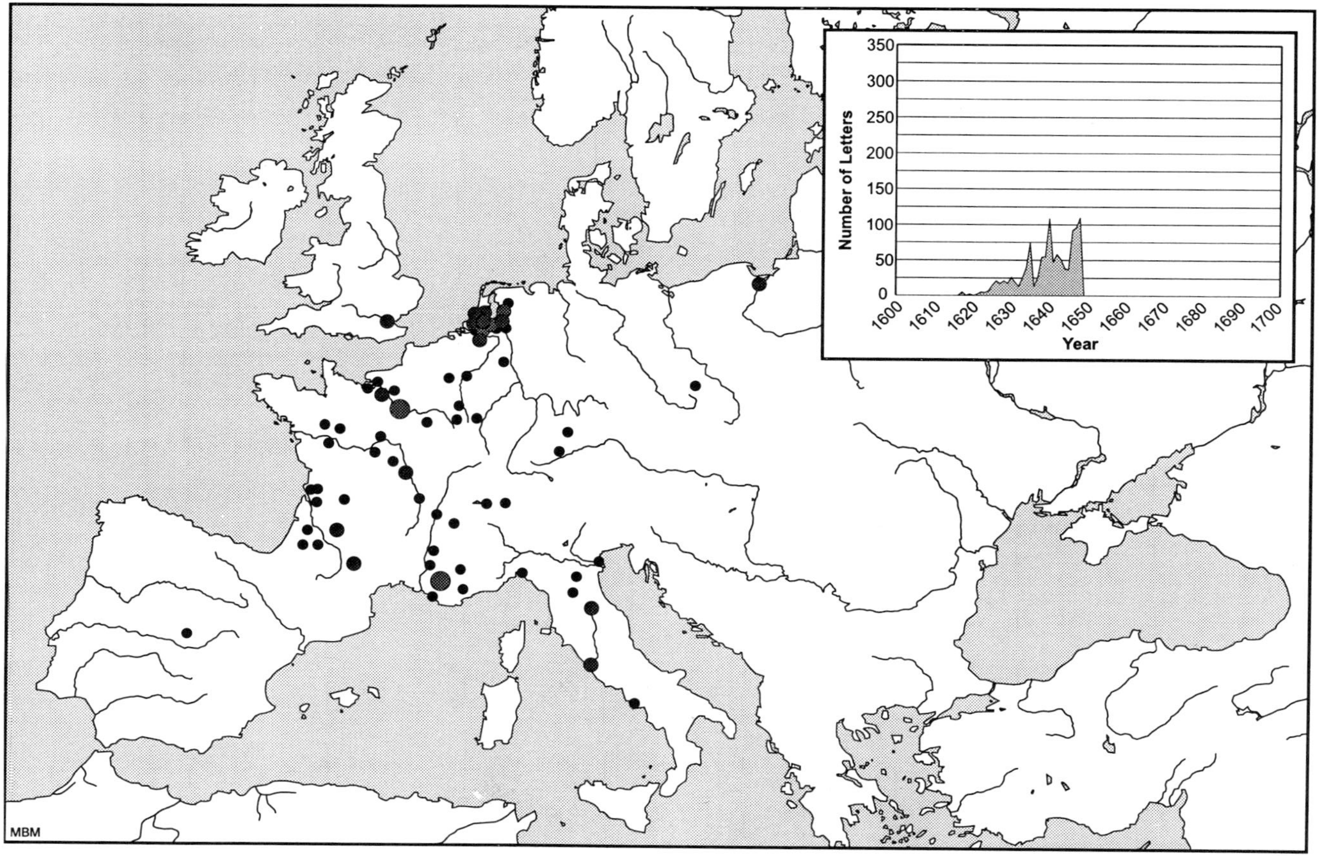

Figure 2. *The Published Correspondence of Mersenne*: The *Correspondance du P. Marin Mersenne* (17 vols.) contains 1,871 entries; slightly over half involve letters to or from Marin Mersenne (1588–1648). The present map and chart represent 1,099 published letters of Mersenne for the years 1617–48.

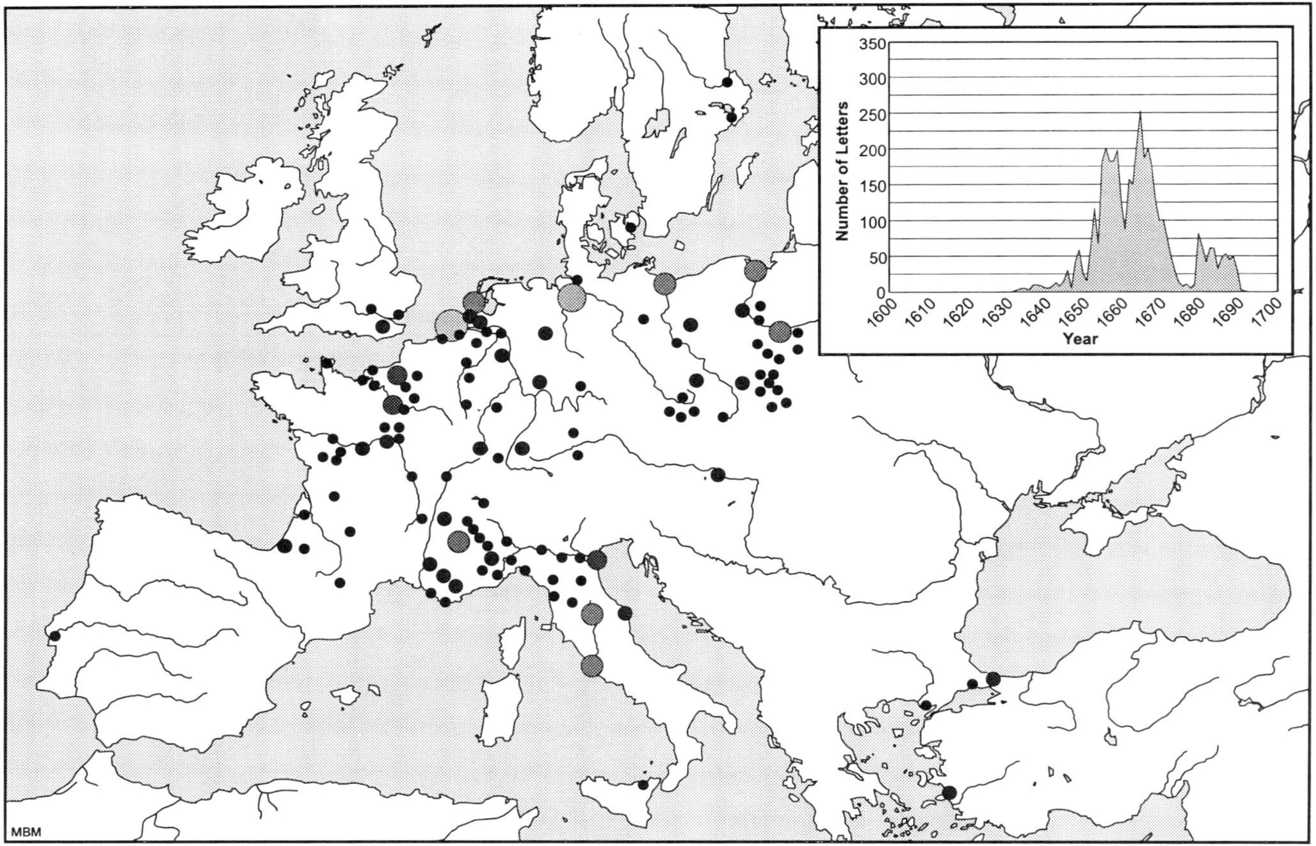

Figure 3. *The Correspondence of the Collection Boulliau*: Ismaël Boulliau (1605–94) left an estimated 10,000 letters at his death; about half that number is extant. The present map and chart represent some ,4200 extant letters from the *Collection Boulliau* for the years 1632–92.

The second major intelligencer, Marin Mersenne, has been called the 'letter box' of Europe [Fig. 2]. Since its inception over sixty years ago, the published *Correspondance du P. Marin Mersenne* (16 vols., 1932–86) has helped to provide a clearer picture of science and learning in the first half of the century. Although Mersenne is traditionally considered one of the most prolific correspondents of the Scientific Revolution, his network was surprisingly small. A survey of his correspondence shows only some 1,100 letters. Well-known to specialists, an identifying feature of his edition are numerous letters (or more often extracts) neither to nor from Mersenne.[9] What is clear is that the Minim's network supplies a critical record of learned activity during the second quarter of the seventeenth century (1617–48), particularly his relations with Peiresc, Gassendi, and Descartes. The geographical distribution of Mersenne's exchanges suggests substantial contact in France, Holland, and Italy.

A longtime friend and colleague of Mersenne, Boulliau lived for some time with the Brothers Dupuy. It is likely that he assumed the epistolary habit by their example. Embracing the humanist ideal of community and communication, Boulliau established a decidedly scientific and European network. To be sure, a vast portion of his letters are political and diplomatic in character, and many reflect his interests in history, classical studies, and theology. Most noticeable is the size of Boulliau's correspondence, which rivals the combined efforts of Mersenne and Oldenburg [Fig. 3]. As the illustrations suggest, Boulliau's correspondence network included some 4,200 letters for the years 1632–93; letters outside the *Collection Boulliau* bring the number to over 5,000. Aside from the size of the network, it should be noted that Boulliau's correspondence fills the chronological gap between the published letters of Mersenne and Oldenburg; it also marks a critical transition in geographical distribution, which now extended beyond France, Holland, and Italy, to Poland, Scandinavia, and the Levant.

The English 'clearing house' of science, Henry Oldenburg was the hurried but capable right hand of the Royal Society. Representing the fourth generation of intelligencers, the *Correspondence of Henry Oldenburg* (13 vols, 1975–86) serves as a diary of late seventeenth-century science, the actualisation of Hartlib's hopes for a universal 'office of address'.[10] As the editors have argued, Oldenburg was the first professional scientific administrator – indeed, his correspondence must also be understood in context with his other productions, notably the *Philosophical Transactions* – and comparable industry and influence is not to be found among the French, Germans, or Italians during his lifetime. The Oldenburg correspondence represents some 3,100 letters over the period 1641–77 [Fig. 4]. Most notable, though no less predictable, is the surge of activity that accompanies the societies' growth. Although England is very well represented, Society contacts in France, Holland, Italy, not to mention the Levant and the Colonies, signal a different sort of enterprise.

9 It may be instructive to note that the first three volumes (MC 1–3) contain more letters involving Peiresc than Mersenne; entry numbers 1–600 involve Peiresc in 217 entries; for 1634, 1635 and 1636 there are 279 entries, 158 involving Peiresc.

10 For an analysis of Samuel Hartlib's archive, see Mark Greengrass's excellent chapter, above.

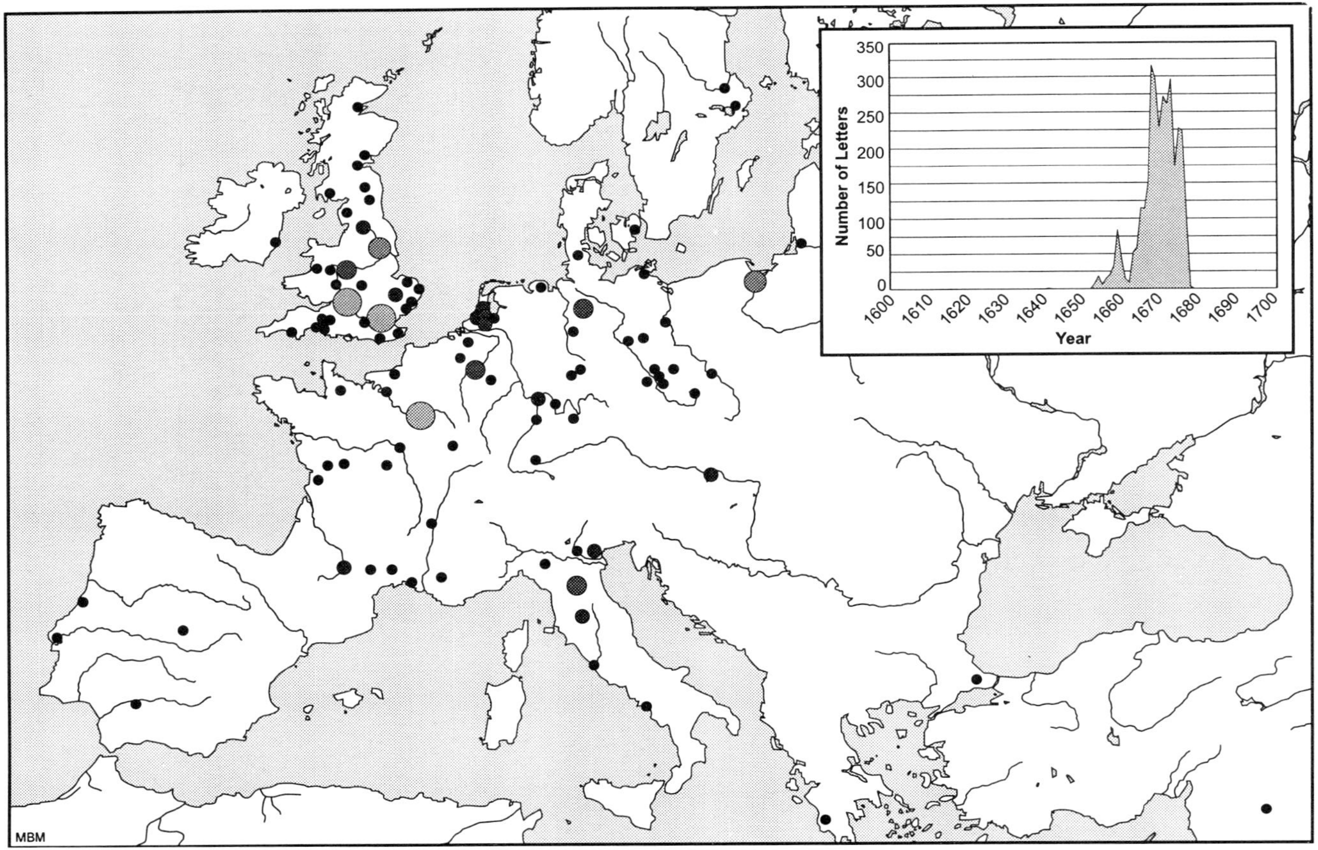

Figure 4. *The Published Correspondence of Oldenburg*: The *Correspondence of Henry Oldenburg* (13 vols.) includes letters to and from Henry Oldenburg (1619?–77) as well as entries for known letters no longer extant. The present map and chart represent 3,170 letter entries for the years 1641–78.

The Archive Boulliau

Although Boulliau was an active member in the Republic of Letters and was included among 'the illustrious of the century',[11] his archive was not kept intact. Shortly after his death his entire library – books, manuscripts, correspondence – was dispersed. The sale of the *Archive Boulliau* set the stage for a century of dispersal and division followed by dramatic episodes of forgery and theft involving Guglielmo Libri (1802–69) and the 'prince of forgers', Denis Vrain-Lucas (1818–?).[12]

Although little is known of the history of the *Archive Boulliau*, evidence suggests that the Bibliothèque du Roi acquired a major portion of the Boulliau papers sometime before 1782.[13] Thereafter, when the Bibliothèque Nationale reorganised its holdings in 1862, the acquisition was divided and reclassified to form the present-day *Collection Boulliau* (BN Paris, f.fr. 13019–13059; forty-one volumes, 23,000 pages). Materials outside the *Collection Boulliau* include over 7000 pages of manuscripts now found in some forty-five archives in a dozen countries.[14]

[11] For contemporary assessments of Boulliau's contributions, see the eulogy of Louis Cousin, *Journal des Sçavans* (Tome 23, 1695); Charles Perrault, *Les Hommes illustres qui ont paru en France pendant le XVII*[e] siècle (2 vols., Paris, 1696–1700). A medal was struck in Boulliau's honour in 1723; see Henri Fatio, *Les Jetons représentant les hommes illustres du siècle de Louis XIV, Gravés en 1732, par Jean Dassier* (Geneva, 1913), pp. 25–6.

[12] Libri's manuscript thefts included entire volumes as well as individual manuscripts and letters, and were nuanced by forgery, falsification, and defacing manuscripts. Exposed by Delisle, Libri was fond of the manuscripts of Boulliau, Dupuy, the family Godefroy, Racine, Rubens, Peiresc, and particularly Hevelius. Here alone, 78 of 110 letters exchanged between Hevelius and Boulliau were stolen. From the *Archive Boulliau*, Libri stole items involving dozens of correspondents, including Hevelius, Gassendi, Blaeu, Fermat, Heinsius, Marucelli, and Seth Ward. See Léopold Delisle, *Catalogue des manuscrits des fonds Libri et Barrois* (Paris, 1888), and Ludovic Lalanne and Henri Bordier, *Dictionnaire de pièces autographs volées aux bibliothèques publiques de la France* (Paris, 1851). Two decades after the 'Affaire Libri' insult followed injury with a second 'Affaire' named after Denis Vrain-Lucas. This hoax touched – or reflected – the *Archive Boulliau* as pointedly as any other. Forgeries include some 95 'letters' of Boulliau to Louis XIV, Huygens, Flamsteed, and Pascal (not including extracts). There were 365 'letters' addressed to Boulliau by Louis XIV (270), Galileo (47), and Pascal (22). Michel Chasles (1793–1880) unwittingly purchased forgeries totaling some 27,000 manuscripts. See Henri Bordier and E. Mabille, *Une Fabrique de faux Autographes ou récit de l'Affaire Vrain Lucas* (Paris, 1870).

[13] There are few printed references to the Boulliau papers, and most quote each other. The earliest is N.-T. Le Prince, *Essai historique sur la Bibliothèque du Roi* (Paris, 1782), which mentions 'Un recueil de plus de 80 vol.' (p. 233). Delambre was the first to suggest that Boulliau papers were discovered in Lyon around 1780: see below, n. 47. In turn, Delisle quoted Le Prince and Delambre (Léopold Delisle, *Le Cabinet des manuscrits de la Bibliothèque Impériale* (2 vols., Paris, 1868–81), i, 549) and eventually added further confusion: 'C'est peut-être aussi à la mission de Chardon de la Rochette que nous sommes redevables d'une notable partie de la correspondance de Peiresc.' Here Delisle quotes Le Prince. See *Cabinet des manuscrits*, ii, 18 n. 3.

[14] For a folio by folio inventory of the Collection Boulliau, see Robert A. Hatch, *The Collection Boulliau (FF. 13019–13059): An Inventory* (Philadelphia, 1982). The *Correspondence Boulliau* (forthcoming) is a comprehensive calendar of Boulliau's complete correspondence. For a detailed listing of Boulliau manuscripts see Hatch, *Collection Boulliau*, pp. 565–76. Major repositories include a mélange of Boulliau letters in Österreichische Nationalbibliothek, Vienna, Briefsammlung Cod. 7049–7050; letters to J.G. Graevius, Copenhagen Det Kongelige Bibliotek, MS Thott 1259; various letters to Oldenburg, London, British Library and Royal Society Library; a volume of letters to Dupuy was found in Boulliau's

Reconstructing the *Archive* and correspondence network of Boulliau has involved identifying extant manuscripts and following the 'logic of the letter'. Having prepared an inventory of the *Collection Boulliau* and a chronological calendar of his complete correspondence, a fair picture of Boulliau's original archive has emerged. Although the negative is difficult to prove, evidence suggests that thousands of Boulliau manuscripts are lost. It is not likely that the precise content and arrangement of the original *Archive Boulliau* will ever be known.

Part of the difficulty is due to the dispersal of his library. Boulliau's last known testament (20 August 1691), for example, contains no mention of his manuscripts, and there is evidence that at least part of library may have been dispersed before his death.[15] Happily, these difficulties are balanced by two important documents: the 'Long Inventory' (nothing less than a manuscript catalogue of the *Archive Boulliau*)[16] and a more concise companion, the 'Short Inventory' [1694], a sale list – complete with appraised value and selling price – of Boulliau's library.[17] These documents, in conjunction with the inventory and calendar, are critical elements in reconstructing the *Archive Boulliau*.

What did the primal *Archive* contain? Surely the best known manuscripts are letters Boulliau exchanged with major figures of his day – Galileo, Huygens, Mersenne, and Oldenburg.[18] Less well-known are published letters exchanged

estate and was later noted by Le Prince and La Porte du Theil (Collection Dupuy 18). The vast majority of Boulliau's astronomical manuscripts are conserved at the Bibliotheque de l'Observatoire (hereafter BO), Paris, MS B3 11, MS B5 11–12; letters to Hevelius can be found here in the Epistolae Hevelium, MS C1 1–16; fragments of the correspondence are found in BN, N.a.L. 1642, Lat. 10347–10349 (Copies); other letters are scattered at other archives; letters to J.-H. Boecler, Staats und Universitätsbibliothek, Hamburg, esp. MS Sup. Ep. 1 and 36; letters to Nicolaas Heinsius and Christiaan Huygens, Bibliotheek der Rijsuniversiteit, Leiden, principally the Burman collections and Hugenius 45; letters to C.-A. de Gondi, Archivio di Stato, Florence (Archivio Mediceo); letters to Galileo, Prince Leopold, and Viviani, Biblioteca Nazionale Centrale, Florence, MS Gal. 87–312.

15 It appears that at least one volume of Boulliau's personal manuscripts had been given or sold to P.-D. Huet at least two years before Boulliau's death. A volume at the British Library, 'Ismaelis Bullialdi adversaria literaria', 248 folios (MS Add, 16912). bears a printed seal, dated 1692, of Pierre-Daniel Huet (1630–1721), Bishop of Avranches and longtime friend of Boulliau: 'Ex libris Bibliothecæ quam Illus=/ trissimus Ecclesiæ Princips. D./ PETRUS DANIEL HUETIUS/ Episc. Abrincensis Domui Professæ/ Paris. PP. Soc. Jesu Integra[ ] vivens donavit/ An. 1692'. The volume is described as 'Recueil/ d'un grand/ nombre/ d'anciennes/ Pieces/ en Grec, Latin/ Français &c/ MS' and was 'Purchased of Tho. Rodd/ 17th Apr. 1847' by the British Library. We know its cost to the shilling but its provenance is obscure.

16 The 'Long Inventory' (BN, f.fr. 13051) consists of 69 folios; it is identified (in two different hands) as «Recueil des pieces trouveés/ Dans les manuscrits de Mr Bouilliaud/ [Lequel avoit/ este secretaire de Mr de Thou dans ses ambassades et est au nombre de homes illustres de ce siècle]» (fol. 1). The document is written on both sides of uniform folio paper (c. 24 x 34 cm.); except for addenda near the end of the volume, the initial catalogue is written in a consistent hand. Internal evidence (spelling, grammar, usage) suggests a young copyist or secretary working with an assistant.

17 AN, Minutier Central, MC XII 226, 29/12 [1694], especially fols. 17–21, which is a sale list of historical manuscripts and books. The final item, discussed below, was appraised at and sold for thirty livres: 'Item, cinquante volumes tant in folio qu'octavo de manuscrits touchant les mathematiques et autre matieres prisé trente livres, dans cette matière, 30 livres'. For a brief survey of the contents see E. Zilberbogen-Chapdelaine, *'Le testament de Boulliau', IIIe Congrès national des Sociétés savantes* (Poitiers, 1986), *histoire des sciences*, pp. 21–33. For research purposes I have transcribed and translated the entire document; to my knowledge the 'Short Inventory' remains unpublished.

18 See above, n. 1. See also Marin Mersenne, *Correspondance du P. Marin Mersenne*, ed. P. Tannery, C. de Waard, and A. Beaulieu (16 vols., Paris, 1932–86); Oldenburg, *Correspondence*.

with Pierre Desnoyers, Fermat, Gassendi, Nicolaas Heinsius, Prince Leopold, François Luillier, Gabriel Naudé, and a handful of others.[19] But most of Boulliau's correspondence and manuscripts remain unpublished. Most notable are major epistolary exchanges with Hevelius, Gassendi, Prince Leopold and over 200 other correspondents between 1632–93.[20] To this correspondence can be added a number of unpublished treatises and copies of letters.[21] But it must be underscored that in addition to his own manuscript materials, Boulliau consciously contributed to the 'Archives of the Scientific Revolution'. Trained as an historian and librarian, active in traditional pursuits of the humanist, Boulliau was a tireless collector and copyist.[22] Among his papers are endless notes and extracts of rare manuscripts, and not least, copies of letters of such contemporary and historical figures as Tycho Brahe.[23] Boulliau also preserved the manuscripts of others. Perhaps chief among them, ten volumes of original autographs addressed to Peiresc.[24]

Archive Evidence: Peiresc to Boulliau

> It is confidently reported that M. Peiresc's heirs, for an entire winter, employed the papers they found in his study as kindling to warm themselves. Better had they burned cedar or the wood of aloes. Nature

19 The most substantial is Pierre Desnoyers, *Lettres de Pierre Des Noyers, secrétaire de la reine de Pologne, Marie-Louise de Gonzague, princesse de Montaue et de Nevers, pour servir à l'histoire de Pologne et de Suède de 1655 à 1659 (Lettres de Pierre Des Noyers à Ismael Boulliaud)* (Berlin, 1859). For other published letters, see the bibliography in Hatch, *Collection Boulliau* (n. 14).

20 Hevelius and Boulliau exchanged over 225 letters between 1648–86. Principal manuscript locations are BN, f.fr. 13026, 13043–13044; N.a.L. 1642. The major repository is the BO, MS C1, Epistolae Hevelium, 1–16; many copies are found in the BN, Lat. 10347–10349. Gassendi and Boulliau exchanged over fifty letters (1633–54); though widely dispersed, most originals are in various collections at the BN. Leopold and Boulliau exchanged nearly 100 letters (1649–72) now located at the BN and the BNC Florence. I have transcribed the above exchanges for a future edition.

21 In addition to perhaps a dozen unpublished treatises and thousands of folios of manuscript notes, published treatises include those of Paolo del Buono, Huygens, Michelangelo Ricci and others. For an indication of the astronomical activities recorded in the Boulliau papers, see Alexandre-Guy Pingré, *Annales célestes du dix-septième siècle*, ed. G. Bigourdan (Paris, 1901).

22 For Boulliau's contributions as a librarian, see Henry Harrisee, *Le Président de Thou et ses descendants leur Célèbre Bibliothèque* (Paris, 1905). For copies of letters and manuscripts made by Boulliau, see Hatch, *Collection Boulliau* (n. 14), index, pp. 473–561; copies of letters include Queen Christina, Maria Cunitz, Paolo del Buono, Flamsteed, Galileo, Gassendi, Hevelius, Huygens, Leopold, Lubienietzki, Melanchthon, Peucer, Portner, Ericus Puteanus, Ricci, Riccioli, Rinaldini, Madame Saumaise, A-M van Schurman, Viviani, and others.

23 See *Tychonis Brahe Dani Opera Omnia*, ed. J.L.E. Dreyer (15 vols., Copenhagen, 1913–29), particularly vols. 6 and 8. The Boulliau copies, involving one letter and two association letters of Tycho, are found in the *Collection Boulliau*, f.fr. 13040. It is possible that Boulliau made copies of the letters while visiting Hevelius in 1661. The Danzig astronomer owned not only Kepler's manuscripts but many of Tycho, now apparently lost. Because Kepler's manuscripts were sturdily bound, they were easily thrown from the window of Hevelius' observatory, which burned to the ground on the night of 26/27 September 1679.

24 In addition to the autographs to Peiresc, Boulliau became custodian for various manuscripts of Gassendi and, little known, François Guyet (1575–1655), the prominent philologist; the 'Long Inventory' announced: '23 caiers contenant divers receuils de feu m^r guiet touchant les origines de la langue latine quil devine de la greque donne par son heritier a m^r iaque depuis lan 1655'.

produces enough of these each day, but such a man as M. Peiresc the world may never again see.

J.-P. de Tournefort, 1717.[25]

Following his death in 1637, Peiresc's celebrated library, which contained ancient manuscripts as well as his personal papers, was placed on offer in Aix and later in Paris. With the assistance of his librarian, Gabriel Naudé, Mazarin purchased many of the ancient manuscripts, and shortly thereafter Pierre Dupuy sorted, arranged, and catalogued Peiresc's private manuscripts. Curiously – and not without importance – it appears Dupuy did not make a separate inventory of Peiresc's letters. At first blush this may not seem unusual. But the absence of a catalogue – or at least brief mention of 10,000–14,000 letters – is out of character. More important, the absence of a catalogue of Peiresc's letters has resulted in the stuff of legend. Circumstances are curious. In his biography of Peiresc, Gassendi insisted that his patron's letters 'do in every way deserve to be published in print'.[26] Close friends with both Peiresc and Gassendi, the Brothers Dupuy were hardly lacking in historical sensibility; directing the daily operations of the Bibliothèque du Roi, they were internationally known humanists and consummate – if not compulsive – cataloguers. Moreover, they were not lacking in scholarly stamina; Pierre Dupuy not only catalogued Peiresc's personal papers, he physically reorganised them – the equivalent of '82 volumes'. The Peiresc letters would have required less effort. Before his death Peiresc personally arranged his correspondence into alphabetised bundles.[27]

But there is no trace of a catalogue.[28] We are thus left to speculate about the most important part of the Peiresc archive. Perhaps, as Francis Gravit suggests, Peiresc's nephew did not consider the letters of value and never sent them to Paris; perhaps Gassendi played a role, having examined Peiresc's letters with great care

25 Joseph Pitton de Tournefort, *Relation d'un voyage du Levant* (2 vols., Lyon, 1717), i, 6: 'Mr de Thomassin Mazaugues est un autre Conseiller du Parlement de Provence, d'un mérite distingué, qui nous fait esperer un recueil des lettres de Mr de Peiresc, dont les manuscrits ont esté répandus par tout le Royaume. Cet homme infatigable en a laissé plus de cent [sic], tous écrits de sa main, comme le remarque Mr Spon. On asseure que les heritiers de Mr de Peiresc, s'étoient chauffez pendant tout un hiver des papiers qu'on avoit trouvez dan son cabinet. N'auroient-ils pas mieux fait de bruler du bois de Cedre ou du bois d'Aloës. La nature en produit tous les jours, & peut-être ne verra-t-on jamais d'homme semblable à Mr de Peiresc.'

26 Gassendi wrote of Peiresc: 'Fearing it would happen as it did, and wishing all not be lost, he followed the course mentioned above, namely, to seize and seek every occasion to include his principal interests in his letters. And hence, I am assured that many of his letters, filled with learning, are the equivalent of as many books, and do in every way deserve to be published.' Pierre Gassendi, *De Vita Peireskij*, Book 6, in *Opera omnia* (6 vols., Lyon, 1658), v, 340.

27 Gassendi wrote that Peiresc was 'was accustomed to digest and bind into bundles with paper, or other covering, all other things, and would write titles in his own hand on each bundle to indicate the contents. In similar fashion, he was accustomed to tie into bundles letters he received, arranging them according to person, place, or time. He first wrote on each [letter] the writer, location, year, month, and day; he eventually added a brief index of the principal topics, which in reading [each letter] he [customarily] underscored': loc. cit.

28 The Dupuy catalogue of Peiresc's personal papers makes no mention of Peiresc's correspondence; the only reference to correspondence concerns two specially-bound volumes of letters of 'illustrious contemporaries', few of which were exchanged with Peiresc.

in writing his biography.[29] What is clear is that the absent list – if such a thing existed – was soon replaced by speculation.

Indeed, over the next two centuries a legend emerged – born from gaps in Peiresc's extant correspondence – and soon grew to heroic proportions. Because no known catalogue of the Peiresc letters existed prior to 1739, it has long been feared that many were lost in the century following Peiresc's death.[30] There is good reason for concern. A striking disparity exists in the number of letters written by Peiresc and the much smaller number of extant letters sent to him.[31] Since the beginning of the eighteenth century, fingers of blame have been directed at a distant heir, Suzanne de Fabri. By various accounts, rather than burn cedar or aloe, she set Peiresc's letters to use in kindling fires, making nests for silkworms, and transforming them into 'paper twists' – 'butterflies'.[32]

Although the legend provided comfort for several centuries, serious study of the discrepancy was not undertaken until some fifty years ago when Francis Gravit published *The Peiresc Papers*. In many ways it was a model study of the provenance of a manuscript collection. In fine, Gravit argued that the legendary discrepancy was not the result of tragic whimsy but archival reclassification, that in fact, a major portion of the long-lost letters was never really lost at all.[33] Ten volumes of autograph letters sent to Peiresc – never mentioned in early catalogues – had in fact been safely stored in the Bibliothèque Nationale since the eighteenth century. Although the existence of the volumes was undeniable – classified as BN Paris, f.fr. 9535–9544 – their provenance had been lost. Lacking a list that declared their first true identity – with no trace of their journey, no map of the existing epistolary landscape – their existence had been lost in legend. Granting it was 'obscure', Gravit gave the volumes a history. The volumes were once owned by Boulliau; they came to the BN with his papers no later than 1782.[34]

29 See F.W. Gravit, *The Peiresc Papers* (University of Michigan Contributions in Modern Philology, no. 14, 1950), pp. 6, 9.

30 The Dupuy Catalogue was published in 1654 by François Henry, one of the editors of Gassendi's *Opera omnia* (Lyon, 1658). Henry is quite clear that he rechecked the Dupuy Catalogue in the presence of the volumes themselves; yet Henry makes no mention of the Peiresc correspondence. The first detailed list appears nearly one hundred years later in the Montfaucon Catalogue of 1739. Although based on the Dupuy Catalogue, it cites for the first time a number of volumes of Peiresc letters (listed as Addenda No. 83 and as Additions 1–11). The so-called 'Boulliau Collection' of ten Peiresc 'volumes' was never listed in any early catalogue.

31 Comparison of the total published letters of Peiresc and other originals, copies, drafts and the registers clearly shows major discrepancies, most notably with the Brothers Dupuy, Cardinal Bagni, F.-A. de Thou, J. Pacius, and F. Luillier.

32 See the epigraph and footnote for this section. Like most legends, elements of the story vary and have been creatively supplemented. See Gravit, *Peiresc Papers* (n. 29), p. 9.

33 As Gravit stated, 'I strongly suspect that the ten volumes of the *fonds français* of the Bibliothèque Nationale (MSS 9535–9544), whose obscure history I shall discuss below (p. 13), account for 1,500 of these 'destroyed' letters, but the loss is still great.' Gravit, *Peiresc Papers* (n. 29), p. 10. The latter study derived from Gravit's dissertation, 'Peiresc: Patron of Scholarship, A study of the influence exercised upon the scholarship of his time by Nicolas Claude Fabri de Peiresc (1580–37)', University of Michigan Ph. D. [1937], Appendix I (pp. 372–446). It is unfortunate that the dissertation has not been published.

34 Gravit follows Delisle; see *Cabinet des manuscrits* (n. 13), ii, 18 n. 3.

Primal Visions: Boulliau and Peiresc

> fifty Volumes, as many in folio as octavo, of manuscripts concerning mathematics and other matters, valued thirty livres
>
> ['Short Inventory', 1694]

Having identified a discrepancy, a legend, and a modern hypothesis to explain an 'obscure history', it is appropriate to take a second look at the 'Boulliau Collection'. Retracing Gravit's steps, we first consider the relation between the Boulliau and Peiresc papers.

Central to the story is the 'Long Inventory' (BN, f.fr. 13051). Consisting of sixty-nine folios, the initial portion of the catalogue (fols. 1–59) is divided alphabetically into separate entries labelled *pieces*. It is a simple list composed largely of printed items, manuscripts extracts and copies – although there are several tantalising paragraphs.[35] The most important entry appears in a separate section that follows the catalogue. On folio 59v we read:

> Letters to M. Peiresc from various illustrious persons from 1600–1650 [sic]. The others are M. Boulliau's, written by him or addressed to him from Rome, Poland, Venice, Constantinople concerning the sciences, affairs of state, which are from the most illustrious persons of the realm, as seen in the summaries. There are others from various persons collected by M. Boulliau concerning astronomy and different subjects. Twenty packets of letters to M. Peiresc from various persons; Twenty two packets of [letters of] M. Boulliau that he has written, received, or collected; Four large packets of mixed letters; Three volumes of bound letters – two of astronomy by the better astronomers, two others of M. Matharel, Resident of the King at Venice during 1648–1651; one volume of letters of le P. Dupuy, Chartreuse at Rome from 1641–1653; another of letters of Prince Leopold of Tuscany with his responses from 1657–1660 – another volume from those of his friends – another of letters of Portner with his responses from 1652–1657.[36]

[35] Fol. 10, for example, provides a list of 'Pieces d'Astronomie' with the note: 'Astronomie divisee en quatre paquets – le ler contient les horoscopes – le 2 contient les problemes le 3me les dissertations mathematiques le quatrieme les observations celestes sans comprendre les lettres concernant ses matieres qui seront dans l'articles des lettres ci-dessous.' Thereafter, the alphabetised *pieces*, previously described as 'vollumes', are abruptly described as 'paquets'; single letter designations are also extended numerically, for example, 'G ler', 'G 2d', 'G 3me', 'G 4me'. As a sample of the contents is given on fol. 11: 'Une infinite d'observations parue manuscrites parues imprimes faites sur l'elevation du pol de plusieurs endroits sur la qualite de l'air sur les comets – sur les planetes, sur les eclipses du solleil et de la lune . . . toutes ses observations ont est faites par M. Bouillau, sont lires des escrits des plus savans astronomes depuis [1520] comme Ticobrahé, Copernic, Gassendi, Kepler, Sheneir, Cujlly, Ritccioli, Bourdin, Hevelius, Longomontanus, Cassiny, et plusiers autres avec quelques lettres de ses messieurs rellies avec les observations.'

[36] 'Lettres a monsieur de peiresc par diverse personnes illustres depuis 1600 iusquen 1650
Les autres sont de M. Bouillaud qu'il a escrit ou qu'on lui a adressé de Rome de Pologne de Venise de Constantinople sur les sciences sur les affaires d'estat elles sont des plus illustres personnes de ces roiaumes comme on le vera par les abregés

On the same folio there is another item of interest. In the top left margin of 59v is a penciled annotation, now barely legible, which translates: 'Packets of letters. Nota bene. [These packets of letters are on deposit at the Bibliothèque Nationale in the present year 1799.]'[37] If the annotation proved authentic, it would establish an earlier acquisition date for the Boulliau papers. Happily, it also pointed in a new direction, to acquisition registers from the eighteenth century.

The *conservateur des manuscrits* at the time was one La Porte du Theil (1742–1815), celebrated for creating what is now known as the *ancien supplément*. Bound in five manuscript volumes, his acquisition records for the late eighteenth century have been carefully preserved (N.a.f. 5427–5431). For present purposes, volume two (N.a.f. 5428) provides the relevant clue, where fol. 99 reads: '430./ CC./ 1. / Trente-cinq Volumes venant de la succession/ de Monsieur Bouillaud. Voyez la notice/ détaillée dans les Bulletins suivans./ 901.' [Thirty-five volumes coming from the succession of M. Boulliau. See the detailed notice in the following note.] Thereafter, between folios 100–105, is a series of virtually verbatim extracts from the 'Long Inventory', and significantly, its marginal annotation is cited. A positive conclusion is reached. The ten volumes of Peiresc letters (numbered 430./CC./3. through 430./CC./12.) were clearly an integral part of the Boulliau papers in 1815 and with little doubt, no later than 1799.[38]

In 1820, five years after La Porte du Theil died, the manuscripts he described, now known as the *ancien supplément*, were further subdivided into the *suppléments grec, latin, et français*. From 1820–62 the ten volumes of Peiresc letters (998–1007) continued to be included in the original sequence of Boulliau volumes. According to Gravit, the Boulliau papers were then catalogued as *supplément* 980–1007bis, with the Peiresc volumes listed near the end as 998–1007.[39] But the

Il y en [a?] d'autres de diverses persones ramassées par monsieu Bouillaud sur l'astronomie et sur differentes matieres.
20 paquets de lettres a M. de Peresque de plusieurs persones
22 paquets de M. Bouillaud ou qu'il a escrite ou recues ou ramasees
4 grots paquets de lettres meles
3 vollumes de lettres rellues – deux d'astronome par les meilleurs astronome deux autres du sieur Matarel resident du roy a Venise in 1648–50–51; un volume des lettres du pere Dupuis Chartreuse a Rome depuis 1641 iusquen 53 – un autre des lettres du Prince Leopold de Toscane avec les responces depuis 1657 iusquen 1660 – un autre vollume de celles de ses amis – un autre des lettres de Portener avec les responces depuis 1652 iusquen 1657.' BN, f.fr. 13051, fol. 59v.

37 BN, f.fr. 13051, fol. 59, 'Paquets de Lettres. Nota. [Ces paquets de lettres se trouuent en Dépositaire de la Bibliothèque nationale en la présente anneé 1799.]'

38 The du Theil catalogue offers much more. Fol. 139 goes on to mention vol. 430./CC./34: 'Sur la couverture de ce Volume, en dedans, est écrit, sans aucune connoissance de l'orthographe[:] Cecy a été trouvé à l'inventaire de Mr Boulot (sic), lequel avoit été secretaire d'ambassade de M.r de Thou.' La Porte du Theil then writes: '[dans ce volume, j'ai trouvé la note suivante: Recueils de Mr. Bouillaud, achetés chez Mr. de S.t Port, et qui regardent l'astronomie, l'astrologie, quelques affaires ecclesiastiques, nouvelles et l'histoire du temps./ Deux paquets, 9 Volumes.] A critical clue, ignored by Gravit, concerns 'Mr de St Port', whose name is associated with more curious aspects of the Gassendi correspondence.

39 Gravit twice cites BN, N.a.f. 5497, 'Catalogue du supplément français' but provides no folio reference. Gravit, *Peiresc Papers* (n. 29), p. 14 n. 54; Gravit, 'Peiresc: Patron' (n.33), p. 387 n. 50. I have not been able to confirm the source, though it is in conflict with Henri Omont, *Catalogue général des manuscrits français, ancien supplément français, Tome III, Nos. 13091–15369 du fonds français* (Paris, 1896), pp. 370–1. The Omont catalogue supplies a concordance whereby the present-day Collection Boulliau

critical change occurred in 1862 when the Peiresc volumes were separated from the initial Boulliau acquisition. This reclassification is the ultimate source of confusion about the apparent loss and actual provenance of the Peiresc volumes. For practical purposes, the *Collection Boulliau* (f.fr. 13019–13059) gained its present identity when the Peiresc volumes (f.fr. 9535–9544) were removed (1862).

Gravit's achievement was quite remarkable. In clarifying the obscure history of the Peiresc volumes, he replaced the fireworks of legend with detailed and creative research. He also demonstrated that the discrepancy in the Peiresc letters was not so large as feared. For all of that, Gravit recognised that a substantial discrepancy remained.[40] But before concluding his account of his 'Boulliau Collection', he made one final attempt to fix an earlier date of acquisition. Following Delisle,[41] he returned to the classic volume of Le Prince (1782). In describing one of the early acquisitions of the Bibliothèque du roi, Le Prince had written:

> 14° A collection of over 80 volumes in folio and quarto of original letters of a number of illustrious men of letters, the celebrated Gassendi, Boulliau, Portner, Spinoza, de Thou, Lomenie, Peiresc, Dupuy, and others. All the letters contain explanations on different points of Greek, Latin, and French literature.[42]

Quoting Le Prince brought several benefits. Gravit not only pushed the acquisition date for the Boulliau papers from 1799 to 1782, he hoped to bring authority and coherence to his argument. But there was a problem. If the Le Prince citation provided a clearer chronology and collateral support, it also introduced an apparent discrepancy. At the close of a concise and otherwise tightly argued case, Gravit concluded:

> I suspect '80' to be a misprint for '30', which would about correspond to the Boulliau papers. The most that can be said definitely is that these letters *may*

aligns with 969^{1} – 997bis and 1007bis. Early commentators refer to 39 volumes, nos. 969–997; see L. Lalanne and H. Bordier, *Dictionnaire de pièces autographs volées aux bibliothèques publiques de la France* (Paris, 1851), pp. 75–9, esp. p. 76.

40 The binding of the Boulliau and Peiresc volumes presents some confusion. Lalanne and Bordier suggest the Boulliau papers had been bound after 1830 (but before 1851) as had the ten volumes of the Peiresc letters (Lalanne and Bordier, *Dictionnaire* (n. 39), pp. 17, 76, 220). But the *Collection Boulliau* (f.fr. 13019–13059) and the Peiresc volumes are bound differently. This suggests that the two collections were considered physically distinct prior to the 1862 reclassification. For all that, volume nine of the Peiresc volumes (f.fr. 9543, not 9542) contains manuscripts of de Thou for the years 1657–61, which confirms the shared provenance of the Boulliau and Peiresc papers. See Lalanne and Bordier, *Dictionnaire*, p. 220 n. 4, and Gravit, *Peiresc Papers* (n. 29), p. 14–15 n. 55.

41 On this issue both cite Prince, *Essai historique* (n. 13) p. 233; Delisle, *Le cabinet de manuscrits* (n. 13), i, 549 n. 2; Gravit, *Peiresc Papers* (n. 29), 15.

42 '14° Un recueil de plus de 80 vol. in-fol. et in-4° de lettres originales de plusieurs hommes illustres dans les lettres, des célèbres Gassendy, Bouillaud, Portener, Spinosa, de Thou, Lomenie, Peyresc, Dupuy et autres. Toutes ces lettres contiennent des éclaircissemens sur différens points de littérature, grecque, latine et françoise.' Le Prince, *Essai historique* (n. 13), p. 223 (Nouvelle Édition, Paris, 1856, p. 192). The orthography is unchanged in subsequent editions.

have been in the Bibliothèque Nationale in 1782, and were certainly there in 1799.[43]

Throughout his analysis Gravit's chief concern was to explain two discrepancies; in the first instance, in the number of extant volumes, and at the close of his study, in his sources. To reach a positive conclusion, the coherence of his argument demanded a consistent number of 'volumes'. But in the end, despite a new level of historical detective work, Gravit's conclusions must be reconsidered. In brief, what follows suggests that his analysis simply replaced one discrepancy for another, that in place of a legend that once read 'kindling, silkworms, and butterflies' we now read 'misprint'.

From Working Habits: Re-Reading Boulliau and Desnoyers

Pierre Desnoyers (1607–93), secretary to the Queen of Poland, Marie Louise de Gonzague, was Boulliau's principal correspondent.[44] With little question, this exchange presents the most complex and potentially rewarding problem of the *Archive Boulliau*. Most noticeable in the correspondence is that half is missing. Boulliau exchanged letters with Desnoyers on a weekly basis for nearly forty years, from 1655 to 1693; his autographs –just under 2,000 originals – have vanished.[45] Happily, most of the Desnoyers letters survive and are currently bound in eight volumes. But curiously, the binding, format, and location of the volumes are not the same – indeed, they are in different archives.[46] Five volumes (bound in folio) are found in the Bibliothèque Nationale (Paris):

f.fr. 13019 (1655–57), 320 folios;
f.fr. 13020 (1657–59), 335 folios;
f.fr. 13021 (1680–83), 352 folios;
f.fr. 13022 (1684–88), 385 folios;
f.fr. 13023 (1689–92), ending with folio 116;

43 Gravit, *Peiresc Papers* (n. 29), p. 15.

44 French by birth, Desnoyers was situated in the Polish court for over forty years. He was a major correspondent of Roberval, intermediary for Hevelius, and was in regular contact with Paolo del Buono, Burattini, and other Italians who populated the Polish court. Good evidence shows Boulliau's letters were read in court to the Queen.

45 There are otherwise unexplained copies of letters of Boulliau to Desnoyers at the Observatoire (several printed by Olhoff) and an interesting 'orphan' at the BN, N.a.f. 6197. To bring perspective to the loss, it is perhaps worth noting that the missing half of this exchange is nearly twice the total number of published Mersenne letters.

46 I have searched widely for Desnoyers manuscripts. Beyond the collections at the BN and AMAE Paris, Desnoyers manuscripts can be found at the BO and AN, and the library at Chantilly, which contains a number of private notebooks. Highly placed in Poland and protected in France, Desnoyers retained a staggering number of manuscripts; it seems inconsistent that Boulliau's letters to him were not carefully conserved. On the basis of his handwriting, I identified several unsigned and undated Desnoyers autographs at the Bodleian (MS Rawlinson C 364); there is no record of acquisition.

while the other three volumes (bound in quarto), Desnoyers' letters for the middle years (1660–73), are found in the Archives du Ministère des Affaires Étrangères (Paris):

Correspondance Politique, Pologne, Tome 14 (1660–65), 552 folios;
Correspondance Politique, Pologne, Tome 25 (1666–69), 499 folios;
Correspondance Politique, Pologne, Tome 37 (1670–73), 349 folios.

Although the importance of this exchange is obvious – particularly given the loss of the Boulliau letters – the Desnoyers volumes betray a long history of confusion. The earliest printed reference comes from the nineteenth century when Delambre reported that three quarto volumes of letters of Desnoyers to Boulliau (for the years 1660–73) were located in the Bibliothèque royale de Paris.[47] Later Delisle, citing Delambre, appeared to nod in agreement, thereby codifying claims that the letters had been discovered in Lyon by Mercier, abbé de St-Léger, and indeed, that the three volumes were under his care at the BN.[48] Given different circumstances I would not disagree with Delisle.[49]

The challenge, of course, is to fix the details of acquisition and unravel the mystery of separation. The place to begin is format, binding, and content. The five folio volumes at the BN appear to have been bound in full calf sometime after 1830; the three quarto volumes at the Archives, which have an early binding embossed with the arms of Colbert, were probably bound in the nineteenth century, during the monarchy of July.

Although more difficult to assess, internal evidence also suggests separate acquisition. The five volumes at the BN, for example, contain a significant number of enclosures and frequent annotations by Boulliau, but usually written on the letters themselves. As a rule, Boulliau was meticulous in dating, annotating, and cross-referencing his correspondence and manuscript library [see Appendix].[50] In

47 'Boulliau était en correspondance avec les gens de lettres les plus célèbres de son temps. On conserve, à la bibliothèque royale de Paris, 3 vol, in-4° de lettres qui lui furent adresées, de 1660 à 1673, par Desnoyers, secrétaire des commandements de Marie de Gonzague, reine de Pologne. C'est à Mercier de St-Léger que l'on doit ces manuscrits, qu'il avait découverts dans un voyage à Lyon.' D-L-E [Delambre], 'Boulliau', *Biographie universelle* (83 vols., Paris, 1811–53), v, 338–40, p. 338; cf. Nouvelle Édition (Paris, n.d.), v, 248–9, p. 248 col. 2.

48 In Vol. I, Chapter 9, 'Acquisitions Diverses, 1775–1789', Delisle cites Delambre; I quote the entire passage: 'On peut placer ver l'année 1780 l'acquisition de la correspondance et d'une partie des papiers d'Ismaël Boulliau (mss. français 13019–13059). Mercier de Saint-Léger les avait découverts dans un voyage à Lyon, et il paraît qu'ils étaient entrés à la bibliothèque du roi avant 1782.' Léopold Delisle, *Le Cabinet des manuscrits* (n. 13), i, 549.

49 There has been confusion in the wake of Delambre. Early on, following Delambre and Delisle, Gravit attempted to forge a link between the 'Lyon volumes' and other absent Peiresc letters. But part of his confusion clearly stemmed from the assumption that there were five (not three) 'Lyon volumes' and that they were located at the BN. See Gravit, 'Peiresc Patron' (n. 33), p. 387. The three quarto volumes were actually found at the monastery of Villeneuve-lès-Soissons and finally deposited with the Archives, then at Versailles.

50 Boulliau characteristically noted lapses in his correspondence with Desnoyers, usually on the appropriate letter itself. For example, just before Desnoyers visited Paris in summer 1682, Boulliau wrote on his

contrast to the BN volumes, the three quarto volumes at the Archives contain a number of separate notes written in Boulliau's hand. These personal 'billets' – neither drafts nor summaries – are interleaved chronologically with Desnoyers' originals. Here, Boulliau's practice seems to have evolved from scratch sheets (where he initially deciphered Desnoyers' coded passages) to more general reminders about topics dispatched to Desnoyers. In any case, no personal 'billets' are found in the BN volumes. Differences in binding, format, and content suggest separate acquisition.

Internal evidence aside, new findings underscore the importance of external evidence. There is now little doubt that the Desnoyers volumes have been separated since the eighteenth century. Circumstances surrounding St-Léger's discovery confirm that three volumes were discovered at the monastery of Villeneuve-lès-Soissons; new documents show St-Léger gave the letters to Condorcet, who then sent them to Versailles where '3 volumes petit in 4° sans être reliés' were accepted for deposit 26 January 1777.[51] In sum, the three quarto volumes never entered the BN; internal and external evidence points to the same conclusion.

One final discovery may speak volumes. Recently revisiting a familiar folio in the Correspondance Politique-Pologne, tome 25, fol. 208, I noticed a half-page fragment written on recto and verso but in different hands. The recto is a summary note by Boulliau dated 22 July 1667; verso is a torn fragment signed and dated from Montpellier, 4 March 1616. Although I did not recognise the verso signature immediately, I soon confirmed it was signed by 'Pacius'. The patient reader may wonder how Boulliau came to write on M. Pacius' letter.

last letter from Amsterdam (8 June 1682): 'Mr Des Noyers arriva a Paris a la fin de Juin' (f.fr. 13021, fol. 231r); on his first letter after departing Paris (Amsterdam, 11 September 1682) Boulliau wrote: 'Mr des Noyers partit de Paris le 29 Aoust po^r retourner en Pologne' (f.fr. 13021, fol. 232r). Several years later, Boulliau explained: 'Monsieur Des Noyers partis de Pologne avec Mr. Morstin et toute sa familla sa femme et son fils au commencement d'Aoust 1685. Et arriverent a la fin du mois. Mr Desnoyers tomba malade d'un rheume qui luy dura long temps. Il passa une grande partie a Paris et arriva a Varsovie le 15/15 Mars 1686' (f.fr. 13022, fol. 160v). Although Boulliau noted these breaks in correspondence, Desnoyers' letters for 1674–79 have not been located.

51 Mercier, abbé de Saint Léger (1734–99), librarian of Sainte Geneviève, explained: 'En 1773, je trouvai chez les Célestins de Villeneuve-les-Soissons la correspondance originale de Desnoiers, secrétaire de Marie de Gonzague reine de pologne avec Ismael Boulliaud savant mathématicien, et j'obtins de ces religieux la cession de ce manuscrit précieux en trois volumes in 4°. Je les communiquai à Paris à Condorcet, secrétaire de l'Académie des sciences, pour en extraire ce qui concernait les mathématiques et en particulier l'astronomie, après quoi je donnai ce recueil précieux au dépôt des Affaires Etrangères à Versailles; le ministre de Vergennes m'écrivit une belle lettre pour me remercier de ce présent au nom du Roi.' BN, N.a.f. 4559, p. 36. Quoted in F. Secret, 'Astrologie et alchimie au septieme siècle', *Studi Francesi*, 20 (1976), 463–79, on p. 464. New evidence confirms that the volumes were placed on deposit at the Archives in 1777. The acquisition register offers the following description: 'Depuis le 3 janvier 1660 jusqu'au 18 février 1673 / Lettres de M. Des Noyers, secrétaire des Commandements de la Reine de Pologne Marie de Gonzague à Mr Bouillaut / écrites de pologne avec le chiffre de Mr Desnoyers / 3 volumes petit in 4° sans être reliés / envoyé le tout au Dépôt par M. De Vergennes le 26 janvier 1777'. Archives du Ministère des Affaires Étrangères, [Acquisition Registre] 1661–1806 [1777, folio no. 244]. I wish to thank the conservateur du Patrimoine, Isabelle Nathan, for supplying the text of this Registre.

The answer is that Pacius was a major correspondent of Peiresc. The fragment therefore has two sides. On the face of it, Boulliau's note – combined with the fragment – brings closure to longstanding speculation. It can no longer be doubted that Boulliau was in possession of at least a portion of the autograph letters sent to Peiresc – indeed, Boulliau supplies a terminal date of the acquisition (1667). The other side of the story – the context of the Pacius letter – is less certain, as we have only a fragment of a larger picture.

The facts of the matter seem clear. As it happens, Peiresc corresponded with two members of the Pacius family, father and son. The handwriting, date, and location indicate the sender was Jules Pacius (1540–1635), Peiresc's former law professor, long-time friend and correspondent.[52] But where does the fragment fit? Did other Pacius letters meet a similar fate? Circumstances suggest the following. First, by tradition, the 'Boulliau Collection' has long been recognised as the chief repository of extant Pacius originals sent to Peiresc; second, by tradition, the Pacius exchange represents one of the largest discrepancies in the Peiresc correspondence – that is, we have far more letters from Peiresc than from Pacius. The loss is great.

As this essay goes to press, I have not found a match to the Pacius fragment in the two most likely repositories, that is, in Paris[53] or Carpentras.[54] Unfortunately, progress in dispelling the mystery has been slowed by the absence of a list, as no comprehensive calendar of the Peiresc letters exists.[55] Until such a context is established, the known facts surrounding the fragment remain disconnected, and a satisfying solution will depend on luck. Perhaps the fragment – or renewed interest in discrepancies – will kindle interest in the undertaking.[56]

52 There is a large discrepancy in the number of surviving letters sent and received between Peiresc and Jules Pacius or Giulio Pace (1540–1635). See Ch. Révillout, *Le jurisconsulte Jules Pacius de Beriga avant son établissement à Montpellier (1550–1602) d'après un document inédit* (Montpellier, 1882); and Philippe Tamizey de Larroque, *Jules Pacius de Bériga. Compte-rendu du Mémoire de M. Ch. Révillout, avec addition de Documents inédits* (Paris, 1883).

53 Two volumes contain Pacius letters, f.fr. 9538 (letters of J. Pacius to Peiresc) and f.fr. 9541 (letters of J. Pacius and Cl. Pacius to Peiresc). The likely location is 9538, between fol. 152 (J. Pacius, Montpellier, 18 December 1615) and fol. 155 (J. Pacius, Valence, 25 October 1616. Other J. Pacius materials are found in f.fr. 2575 and f.fr. 16,733.

54 The Pacius letters at Carpentras are for 1608, 1614, 1619, and 1630. I wish to thank Anne Favre-Reinbold and Jane Tolbert, Peiresc specialists, for confirming this information. See C.-G.-A. Lambert, *Catalogue descriptif et raisonné des manuscrits de la bibliothèque de Carpentras* (3 vols., Carpentras, 1862–80), iii, and Duhamel and Liabastre, *Catalogue Général des manuscrits des Bibliothèques publiques de France, Départments, Tome 35, Carpentras* (Paris, 1899).

55 It appears that P. Tamizey de Larroque developed, if not a calendar, at least a working list; it could not be located following his death. Omissions in the various published volumes of Peiresc letters suggest, however, that if a calendar existed it was very incomplete. The recently discovered fragment highlights the absence of a Peiresc calendar.

56 If the existence of the fragment is outrageous it also appears, to my knowledge, unique in the context of some 30,000 pages of Boulliau manuscripts.

Concluding Counter Inferences

Tolstoy was by nature a fox, but believed in being a hedgehog.
Isaiah Berlin, *The Hedgehog and the Fox*

Unfortunately, major discrepancies in the reported size of the *Archive Boulliau* cannot be resolved with inventories and calendars alone. Arguably, more archival research is required – and perhaps more careful reading of familiar documents. By way of conclusion, I offer an alternative reading of discrepancies associated with the *Archive Boulliau.*

Much to his credit – but not without irony – Gravit was the first to call attention to the size of the BN acquisition of Boulliau papers. The earliest reference, the reader will recall, came in 1782 when Le Prince reported the acquisition of 'plus de 80 vol. in-fol. et in 4°'. During the next century, Delisle was the first to suggest that the reference was to the 'Boulliau papers' – that is, that Le Prince had noted the first (at any rate primary) acquisition of Boulliau manuscripts by the Bibliothèque du Roi. In this century, in order to account for an apparent discrepancy in his sources, Gravit discounted the number from '80' to '30'. By way of conclusion, I would argue that the discrepancy is genuine, the loss of manuscripts real. Unfortunately, given present circumstances, evidence for the loss is not based on the manuscripts but a consistent reading of sources, specifically, Le Prince ('80' volumes), La Porte du Theil ('Trente-cinq Volumes'), and the two 'Inventories'.[57]

The key source is the 'Long Inventory'. As it happens, this manuscript was well-known to La Porte du Theil, Gravit, and possibly to Le Prince. As described above, it contains an alphabetical list of separate entries or *pieces* described as *volumes* (itemised A–H) and *paquets* (I–Z).[58] Not counting supplements, there are twenty-six such entries. In addition to these listed volumes, as we have mentioned, there is also a second listing that announces 'Plusieurs paquets de lettres': '20 paquets' of letters to Peiresc; '22 paquets' of Boulliau letters, '4 grots paquets' of miscellaneous letters, and a series of bound letters, which appear to come to eleven volumes.[59]

57 Gravit seems to understand 'Trente-cinq Volumes' in the following context. In 1820 the supplement was divided resulting in reclassification of the volumes as 980–1007 and 1007 bis. As is often the case, a given number can be further subdivided with superscript numbers as well as 'bis', 'ter', and so forth. In this case, the reckoning shows the supplement reclassification involved $988^{1,2,3,4}$, as well as 997 bis and 1007 bis. The total number thus comes to thirty three. What is not clear, in the absence of additional archival evidence, is why Gravit ignored the ancien numbers 969^{1}–979^{2}, an additional volumes (eventually reclassified as f.fr. 13019–13036), which would bring the total to 51 volumes. A further difficulty is the number of Peiresc 'volumes'. Both La Porte du Theil and Gravit were familiar with the 'Long Inventory', but because the letters to Peiresc had been bound by the time of du Theil's report, the '20 paquets' described in the 'Long Inventory' had been reduced to ten bound volumes. But even here, given a reduction from '20 paquets' to '10 volumes', the result would be 47 volumes.

58 There is one fewer letter in this alphabet, as letters 'I' and 'J' were traditionally substituted; the packet marked 'Numero 9' probably resulted from the scribe's assistant misreading the letter 'Q'. The catalogue also includes addenda (fols. 57v–59, 65–66v, 67–69v). Written on both sides (fols. 1–59) the list contains 1,860 items on 117 folio pages.

59 Following the long alphabetical list is a summary description of an entirely different group of

In sum, as an alternative reading of earlier reports, it appears that the first published notice concerning of the 'Boulliau papers' cannot be dismissed or simply down-sized in the name of convenience. If my reading is correct, the 'Long Inventory' of the *Archive Boulliau* represents some '83 unit entries', which would about correspond to the Boulliau papers reported by Le Prince – 'plus de 80 vol. in-fol. et in-4°'.[60] Hedging will no doubt continue.

Appendix
Boulliau's Organisational and Access Procedures

In the course of studying Bouillau's manuscripts, substantial detail has become available concerning his methods of handling his archive. In view of its significance for Bouillau's intellectual personality, and by way of comparison with other figures surveyed elsewhere in this volume (e.g., Michael Hunter's study of Boyle), a brief summary is provided below.

Individual Documents: Annotations

Bouillau gives the title and date of virtually all manuscripts and copies; as a rule, he also dates manuscripts and letters undated by others, corrects dates of others when in error, and records the date of receipt when the letter is undated. When a new exchange is initiated, Boulliau often identified new correspondents with several lines of annotation; with established correspondents and friends, he frequently explains the epistolary lapse or change in frequency by noting the correspondent's change in circumstance, travel activities, and so forth.

Organisation of Letters, Manuscripts and Printed Documents

Evidence suggests that Boulliau arranged incoming correspondence chronologically by person, often including inserts, summaries, extracts, and related manuscript or printed material. It seems clear that he originally grouped and arranged his correspondence and papers in 'bundled paquets' which he labeled by correspondent and year. Copies of related letters and data were generally interleaved chronologically. There is also evidence that his manuscript bundles (paquets) were

manuscripts (fol. 59v); it is written in the same hand as a physical continuation of the catalogue but begins distinctly with a new 'page' and heading.

60 Gravit's enumeration seems to stem from La Porte du Theil (BN, N.a.f. 5428, fol. 99r), itself based on pre-*Ancien Supplément* numbers, which begin with 430/CC./1. Even here individual numbers (430/CC./1–35) sometimes refer to more than one paquet or volume. Multiple volumes indicated by a single number are also evident in the *Ancien Supplément*, for example, 988 represents four volumes denoted by superscript 1–4.

arranged by topic (history; classical studies; philology; horoscopes; astronomical data; theology; and experimental physics). Astronomical data was generally arranged by subject and year, e.g. by planet, by phenomena (lunar eclipses), by object (Mira Ceti). Here, Bouillau displayed little organisational regard for authorship or format (letter, manuscript, pamphlet).

Reference and Access Practices

As a rule, Boulliau made and retained drafts of outgoing letters in Latin, though he did retain copies of 'important' letters regardless of language. Letters, manuscripts, and observational data were frequently copied and re-transmitted.

Boulliau cross-annotated his letters and manuscripts. For example, astronomical, classical, and historical texts are often referenced, annnotated, or physically arranged together. There is ample evidence that astronomical data were often re-assessed, annotated, supplemented, and interleaved with new materials regardless of date (letters, attachments, inserts, printed texts). Similar practices are evident with his large horoscope collection.

Finally, there is wide use of cipher in many of Boulliau's letters, particular those dealing with political and diplomatic issues. Passages written in cipher are never entirely de-coded (usually as annotations on the incoming letter); this seems to suggest that Boulliau had memorised many of the cipher symbols.

5

Ireland as a Laboratory: the Archive of Sir William Petty

Frances Harris

We know Sir William Petty today chiefly as one of the founders of modern economics.[1] But in his own day he was famous for much else besides: one who 'for his Tallents in Philosophy, Physick, Navigation, Poetry, Surveying, Calculations and Politicall Arithmetick was in his Generation most eminent', in the words of his cousin and self-appointed archivist, Sir Robert Southwell.[2] Only days after his death Viscount Weymouth remarked, 'I wish some knowing person might have the perusal of his papers, for I am told he had excellent things by him.'[3] Some did have access to the quantities of writings which Petty left behind in his closet. John Aubrey was one, though he did no more than note the titles of a handful of works.[4] But primarily the archive was regarded as 'a family jewel'.[5] From the eighteenth century onwards it was preserved at Bowood in Wiltshire, the home of Petty's descendants the Earls of Shelburne and Marquesses of Lansdowne, where it was described by Michael Hunter as 'perhaps the most important collection [for the Scientific Revolution] still in private hands'.[6]

Few of Petty's writings were printed in his lifetime, though some had a limited manuscript circulation. Further selections have been published since, but many still remain unpublished in the archive. Until the 1970s, when it was listed and microfilmed, access to it was mediated chiefly by the writings and editions of his descendants.[7] Yet much in the archive, as we shall see, derives from or relates to

I am grateful to Dr Tony Aspromourgos and Dr T.C. Barnard for their valuable comments on drafts of this paper.

1 Guy Routh, *The Origin of Economic Ideas*, 2nd edn (London, 1989), p. 35; Tony Aspromourgos, *On the Origins of Classical Economics* (London, 1996), pp. 2–3.

2 *The Petty-Southwell Correspondence*, ed. 6th Marquess of Lansdowne (London, 1928), p. 333.

3 Weymouth to Southwell, 4 Jan. 1688, B[ritish] L[ibrary] Add. MS 72855, fol. 127v.

4 John Aubrey, *Brief Lives*, ed. Andrew Clark (2 vols., Oxford, 1898), ii, 146.

5 BL Add. MS 72854, fol. 1: note by Abraham Hill, FRS, 1703, on the Petty manuscripts preserved by Southwell.

6 Michael Hunter, *Science and Society in Restoration England* (Cambridge, 1981; reprinted Aldershot, 1992), p. 218.

7 Works drawing on the archive, written or edited by Petty's descendants include Lord Edmond Fitzmaurice, *The Life of Sir William Petty* (London, 1895); and the following, all written or edited by the 6th Marquess of Lansdowne: 'The Lansdowne Maps of the Down Survey', *Proceedings of the Royal Irish Academy*,

the public domain, and its most important single item, the Down Survey of Ireland, was the instrument of Cromwell's Irish policy and remains one of the great state documents of the seventeenth century. In 1993 the whole archive was purchased by the British Library, where it has been incorporated into the manuscript collections as Add. MSS 72850–72908. It has been my task to decide how best to deal with it now that it has a permanent home in the national library, and the present paper derives from this process and the preliminary conclusions about the archive which I have drawn from it.

Petty was born in obscurity, the son of a Hampshire clothworker with no fortune but his aptitude for all aspects of mathematical and mechanical learning. Put to sea as a boy on a merchant ship, he was handicapped by his poor eyesight, but acquired a lifelong passion for shipping, navigation and trade. In the early 1640s he turned his studies to medicine at Utrecht, Leiden and Paris, where he came under Hobbes's influence,[8] and on his return to England in 1646, he was drawn into Samuel Hartlib's orbit. His first publication, *The Advice of W.P. to Mr. Samuel Hartlib for the Advancement of some particular Parts of Learning* (1648), displayed a Baconian passion for experimental knowledge and its applications to human betterment, and advocated various schemes to promote 'Mechanicall Arts and Manufactures'.[9] But for Petty these were not an end in themselves. All his life he was a projector and a contractor as much as a natural philosopher. The most telling of the few documents in his archive for the early period of his life is a list of the means by which he planned to make his fortune: it includes the practice of medicine, publishing accounts of 'secrets and artifices truly but not fully & plainly related', procuring patents for mechanical inventions, using these to draw in investors and obtaining 'honorary pensions from great persons'.[10] This, enlarged by opportunities for public employment and the acquisition of estates, continued to be his agenda thoughout his career.

A doctor of physic and Professor of Anatomy at Oxford before he was thirty, Petty soon sought a wider field and found it in Ireland. In 1652 he crossed to Dublin

35 (1919), 385–407; *The Petty Papers* (2 vols., London, 1927); *The Petty-Southwell Correspondence* (London, 1928); *The Double Bottom or Twin-hulled ship of Sir William Petty* (Oxford, 1931), and *Glanerought and the Petty-Fitzmaurices* (London, 1937). C. H. Hull (ed.), *The Economic Writings of Sir William Petty* (2 vols., Cambridge, 1899), acknowledges assistance from Lord Edmond Fitzmaurice, but does not include any manuscripts from the archive. *The History of the Survey of Ireland commonly called The Down Survey*, ed. T.A. Larcom (Dublin, 1851) is based on a collation of the three existing manuscripts, one of which (Add. MS 72874) is in Petty's archive, and *The Census of Ireland 1659*, ed. Séamus Pender (Dublin, 1939) is an edition of the unique copy in the archive (now Add. MS 72876). A 'Report on the Literary, Personal and Official Correspondence and Papers of Sir William Petty', was compiled by M.D. Slatter and distributed by the Royal Commission on Historical Manuscripts, 1980. This describes the archive under its Bowood numbers, Vols 1–23 and Boxes A–H, 12–17. Although the papers have now been renumbered and to some extent rearranged, any work on the archive must owe a great debt to this meticulously detailed list.

8 Petty's fine autograph copy of Hobbes's 'Opticks', with diagrams, is BL Harl. MS 3360; see also Aubrey, *Brief Lives* (n. 4), ii, 139, and Lindsay Sharp, 'Sir William Petty and Some Aspects of Seventeenth Century Natural Philosophy' (Oxford D.Phil. thesis, 1976), pp. 39–46.

9 W.E. Houghton, 'The History of Trades: its Relation to Seventeenth-century Thought', *Journal of the History of Ideas*, 2 (1941), 35–46.

10 BL Add. MS 72891 (Bowood Vol. 3), fol. 8v; see also Sharp, 'Sir William Petty' (n. 8), p. 55.

as physician-general to Cromwell's army. There he also took on the government contract for surveying the lands confiscated from the defeated Irish Catholics for redistribution to the Cromwellian soldiers and investors. It was a huge organisational task, which he carried through with impressive speed and accuracy: training teams of disbanded soldiers to carry out the field work, and setting up a cartographic office in Dublin, in which their books were converted into parish and barony maps with accompanying schedules.[11] In the redistributions which followed Petty acquired large Irish estates of his own, particularly in Kerry. These made his fortune and they also gave him an experimental field large enough for his ambitions. 'The seate of project and designe must bee there at last', as he told one of his agents;[12] and so it remained until the end of his life.

At the Restoration Petty prudently returned to England for a time, was knighted, became a founder member of the Royal Society and contributed intermittently to its experimental programme. His designs and prototypes for a twin-hulled ship, or 'double bottom' as he called it, were the best known of these projects and for a time attracted the encouragement of the King.[13] But his most seminal work was in the new field of economics and demography, in conjunction initially with the London statistician, John Graunt.[14] Petty's terms for this new science, 'political anatomy' and 'political arithmetic', recalled both his own medical and mathematical studies and Bacon's 'judicious Parallel . . . between the Body Natural, and Body Politick'; Sir Robert Southwell called Petty 'the first man that ever brought Algebra into Human Affairs'.[15] *A Treatise of Taxes and Contributions*, published in 1662, was followed by the privately circulated 'Verbum Sapienti, or Discourse about Taxes and the Value of People' in 1665, 'Political Arithmetick' and 'The Political Anatomy of Ireland' in the 1670s, the 'Quantulumcunque concerning Money' in 1682, and several printed essays on the populations of London, Dublin and other capital cities in the 1680s. But during this period Petty was actually spending the greater part of his time (1663, 1666–73 and 1676–85) in Ireland, supervising the experimental colony, with fisheries, marble quarrying and iron works, which he had established on his remote estate in Kerry.

It seemed obvious to Petty's Royal Society fellows after the Restoration that his awe-inspiring powers of 'ratiocination' should continue to be harnessed to the service of the state. 'If I were a Prince, I should make him my second Counselor at least', John Evelyn commented, 'There is nothing difficult to him.' Aubrey

11 Petty's own accounts are in *The History of . . . The Down Survey* (n. 7); *Reflections upon some Persons and Things in Ireland* (London, 1660); and Fitzmaurice, *Life* (n. 7), pp. 325–9; see also J.H. Andrews, *Plantation Acres* (Ulster Historical Foundation, 1985), pp. 63–71.

12 BL Add. MS 72850 (Bowood Vol. 6), fol. 39: Petty to John Petty, 26 March 1661.

13 See Lansdowne, *Double Bottom* (n. 7), based on papers from the Petty archive, the Pepys Library, Magdalene College, Cambridge, and the Royal Society.

14 For their collaboration over the *Observations on the London Bills of Mortality*, see Hull, *Economic Writings* (n. 7), i, xxxiv–liv, Lansdowne, *Petty-Southwell Correspondence* (n. 7), pp. xxiii–xxxii, and Sir Geoffrey Keynes, *A Bibliography of Sir William Petty* (Oxford, 1971), pp. 75–7. A manuscript of Petty's 'Verbum Sapienti' (1665) inscribed to Graunt has recently come to light; see Bloomsbury Book Auctions sale cat. 286 (10 Oct. 1996), lot 180.

15 Hull, *Economic Writings* (n. 7), i, 129; Lansdowne, *Petty Papers* (n. 7), i, xxix.

wrote of Petty's 'prodigious working witt' and added that he was 'both fitt for, and an honour to, the highest preferment'.[16] Petty thought so too and complained bitterly in retrospect that he had had no public employment of trust or profit after 1659. He did not choose to count a judgeship of the Irish Admiralty court, which he resigned after a few years because 'it gives me no such work as I expected'; although his inaugural speeches are themselves virtuoso disquisitions on shipping and trade.[17]

The failure to achieve public office was not for the want of trying. Petty made repeated proposals, of which the drafts remain in his archive, for a registry of lands, commodities and inhabitants in England and Ireland, initially using the surveys of the Cromwellian period in which he had taken a large part, and employing the hearth tax officials to collect the statistics. Characteristically these proposals were aimed not just at collecting data for his economic theorising, but at reforming the whole tax system and at realising his ambitions as well, since it was his own appointment as surveyor-general or 'accomptant-general' in charge of these operations which he envisaged.[18] But it never came about. This disappointment, which made him reject the compensating offer of a peerage, he attributed to being 'suppresst and opprest' by a combination of enemies, of whom he certainly had many. To his family and Royal Society associates Petty was a man of warm affections, witty and convivial. But in wider affairs he was a different person: complaining, contentious and impossibly vainglorious about all his projects, from his early and not very successful instrument for duplicate writing ('daily and hourly usefull to all sorts of Persons in all places of the world, and that to perpetuity'), to his experiments in shipping ('blessing the whole world to the world's end'), and the political schemes of his last years (excelling 'all I have ever knowne of that kind').[19]

Petty was a brilliant conceptualiser. It has been remarked how often he formulated ideas, almost casually and in passing, which have since become standard in economic thought.[20] But his projects tended to be a good deal less impressive in practice. The public failures of several of the prototypes of his twin-hulled ship did not inspire confidence. Nor, as T.C. Barnard has shown, was the colony on his Kerry estates ever a very convincing showcase for his views about trade and plantation. Petty's Irish interests were the making of him, both financially and intellectually, but since both the titles of his estates and the quitrents payable to the Crown were constantly disputed after the Restoration, they were also to be a

16 *The Diary of Samuel Pepys*, ed. R. Latham and W. Matthews (11 vols., London, 1970–83), v, 27; *The Diary of John Evelyn*, ed. E. De Beer (5 vols., Oxford, 1955), iv, 58–9; Aubrey, *Brief Lives* (n. 4), ii, 142.

17 Petition from Petty to James II, 1686, BL Add. MS 72857 (Bowood Vol. 8), fol. 147; Fitzmaurice, *Life* (n. 7), pp. 245, 250; for his papers and draft speeches as judge, see BL Add. MS 72893 (Bowood Box G) and BL Lansdowne MS 1228, fols. 38–56v.

18 Lansdowne, *Petty Papers* (n. 7), i, 77–103, 178–80.

19 *Advice of W.P. to Mr. Samuel Hartlib* (London, 1648), sig. A3; Lansdowne, *Petty-Southwell Correspondence* (n. 7), pp. 129, 242.

20 Routh, *Origin of Economic Ideas* (n. 1), p. 36.

source of litigation and grievance to him for the rest of his life. These concerns in turn got in the way of his larger ambitions. Charles II, after giving him initial encouragement, found him politically troublesome for this reason and relegated him to the role of 'naval scaramouche' as a punishment. The Duke of Ormonde would not have him on the Irish Privy Council because his endless schemes, argumentativeness and complaints hindered business.[21]

The accession of the Roman Catholic James II in 1685 meant a further threat to the interests of the Protestant landowners in Ireland. On the other hand the new King seemed much more receptive than his brother to Petty's ideas. Having spent most of the preceding twenty years in Ireland, Petty returned to England in 1685 for the last time and embarked on his most prolific period of writing. But even James II must have been taken aback by receiving, along with Petty's wide-ranging projects for reform, what he described as 'Indices and Catalogues of the grosse wrongs' which had been done him for thirty years past, together with suggested remedies and rewards.[22] When he died three years later, none of his schemes had been brought to fruition.

'Whatever becomes of me', Petty wrote in his frustration towards the end of his life, 'I can leave such arguments of 50 yeares art and Industry as will be a credit to my children and friends.'[23] He was referring not least to the papers left behind in his closet.

Even when an archive has been bequeathed as an heirloom and preserved within one family as Petty's was, there will inevitably be disarrangement and losses over successive generations: from division of property, from neglect, from use, from plunder by collectors. Sometimes these losses will be permanent; sometimes they will simply result in dispersal and the preservation of some manuscripts elsewhere, in which case they may be regarded as notionally if not physically still part of the archive. Conversely, there may be unexpected accretions as well. Both of these processes have been at work to determine the present content of Petty's archive.

Petty preserved the anatomy lectures for his Oxford professorship and the records of his medical practice carefully, and in the midst of his Irish difficulties in later life would sometimes regret that he had not been content to follow his original profession.[24] They still remain intact in his archive. But the Down Survey records, which Petty valued as his greatest achievement, were much depleted after his death. In his will he mentions three chests of original maps and field books, two versions of the Survey itself, a chest of distribution books, two of loose papers

21 On this whole subject I am indebted to T.C. Barnard's important article, 'Sir William Petty, Irish Landowner', in H. Lloyd-Jones et al. (eds.), *History and Imagination: Essays in Honour of H.R. Trevor-Roper* (London, 1981), pp. 201–21; see also his 'Sir William Petty as Kerry Ironmaster', *Proceedings of the Royal Irish Academy*, 82 C (1982), 1–32.

22 Lansdowne, *Petty-Southwell Correspondence* (n. 7), p. 224: 31 July 1686; BL Add. MS 72857 (Bowood Vol. 8), fols. 147–8, 164–5.

23 Lansdowne, *Petty-Southwell Correspondence* (n. 7), pp. 211-12.

24 Ibid., pp. 214, 223: 13, 31 July 1686; the medical papers (Bowood Vol. 3) are now BL Add. MSS 72891–2.

and his own manuscript history of the project.[25] Of these only the history and one version of the Survey are still part of his archive, and the latter is incomplete, consisting of only five of the original seven volumes.[26] That including the Kerry survey appears to have been lost as early as 1727 in the course of the continuing lawsuits over the estate. The other version of the Survey, containing a more complete but less ornate sequence of the barony maps, was captured by a French privateer while being transported from Ireland to England in 1707 and is now at the Bibliothèque Nationale.[27] Of the field and distribution books no trace remains, and of the two chests of loose papers, only a few miscellaneous items are left.[28] But since the Down Survey maps deposited by Petty at the Surveyor-General's office in Dublin were almost entirely destroyed by fires which engulfed the Irish records in 1711 and again in 1922, this now constitutes the main archive concerning the Survey, famous (or infamous) as the first exercise in human geography on such a scale to be carried out on scientific principles.[29]

The records relating to Petty's development of, and legal battles over, his model 'plantation' in Kerry have suffered similar losses. His letterbooks and those of his agents are now the chief surviving source; but even these are no longer complete. From the early eighteenth century onwards they descended through a different branch of his family (that of the widow of his eldest son after her remarriage), probably because they were bound volumes and stored separately from the loose papers. One volume found its way into the Irish records and appears to have been destroyed, like the Down Survey records, in 1922. The others came into hands of a succession of collectors, the last being Sir Thomas Phillipps, before being dispersed again at auction in 1911. At this point one volume of Petty's letterbooks, that for 1672–75, and all those of his agents, were bought back by his descendants and restored to their place in his archive; but the remainder of Petty's letterbooks, covering the years 1666–72, 1676–78, 1680 and 1682–83, were acquired by Sir William Osler and are now part of the collection which he bequeathed to McGill University in Montreal.[30]

25 Fitzmaurice, *Life* (n. 7), p. 320.

26 The Down Survey volumes (Bowood Vols. 24–28), with a group of loose maps, are now BL Add. MSS 72868–74. Also in the archive are the original manuscript maps, based on those of the Down Survey, of Petty's atlas of Ireland, *Hiberniae Delineatio* (BL Add. MS 72875) and the Census of 1659 (Bowood Box 12, now BL Add. MS 72876).

27 Lansdowne, 'The Lansdowne Maps of the Down Survey' (n. 7), p. 390; Y.M. Goblet, *La transformation de la géographie politique de l'Irlande* (2 vols., Paris, 1930), ii, 21–6.

28 The loose papers (from Bowood Box F) are now BL Add. MS 72877; Petty's day-book as commissioner for satisfying the army's arrears, 1656–59 was once part of his archive. Like his letterbooks, it came into the hands of Austin Cooper and then Sir Thomas Phillipps in the nineteenth century, and is now BL Add. MS 35102.

29 Peter Barber, 'The Down Survey' (British Library Report, 1992), p. 1.

30 Slatter, 'Report' (n. 7), pp. 113–14, 120, 128, 143, 152, 174; annotated extracts of Sotheby's catalogue of the Phillipps sale 27 April 1911 in BL Add. MS 72905. Petty's letterbook for 1672–75, lot 816 at this sale, was formerly Bowood Vol. 19 and is now BL Add. MS 72858. Some extracts from the other letterbooks (now McGill University Library, Osler MS 7612), made while they were still in the Phillipps collection, are in BL Add. MS 72905, and a few extracts from these have been printed in Fitzmaurice's *Life* (n. 7). One further volume for 1677–79 is noted in Add. MS 72905 as having been in the Public Record Office in Dublin in 1834, and was apparently destroyed in 1922. The letterbooks of Petty's

Very few of Petty's incoming business letters survive, and the comparatively small and scrappy collection of related papers must represent only a tiny fraction of those generated by his estate projects and lawsuits, the latter once 'soe vast that no man understands the whole'.[31] There are, however, some excellent letters to and from Petty's wife during their respective absences, chiefly concerning their estate and financial affairs, and the education of their children (another engrossing but frustrating matter of experiment to Petty). Supplementing them are his instructions to his wife for a journey she made to oversee his Kerry affairs in 1675, and her own journal of these weeks; although these were among a number of items from Petty's archive which strayed into the Lansdowne collection of manuscripts at an early stage and were acquired by the British Museum in 1809.[32] They make a significant addition to the growing body of evidence of women's support of scientific and business activities at this period.

Petty's role in two institutions of the new science, the Royal Society and the Dublin Philosophical Society, have received a good deal of attention from historians of science.[33] It is therefore worth pointing out that only a small proportion of his archive relates directly to these activities. In the case of the Royal Society his role was limited by the fact that he spent far more time in Dublin than in London after the Restoration; in the case of its sister institution, of which he was elected first president in 1684, by his return to England for good a year after it was founded. Again, he appears to have kept few of his incoming letters from fellow men of science. There is a selection from such Royal Society associates as Sir Robert Moray, Dr Robert Wood, Viscount Brouncker, Henry Oldenburg, Sir Peter Pett and Sir Robert Southwell. But with the exception of Southwell (of whom more in due course) and Wood (from whom more than forty letters survive), each is represented in the archive by only few surviving letters,[34] and there are no letters at all from John Pell, Samuel Hartlib, Robert Boyle or Lord Anglesey, although we know from other sources that Petty had intermittent correspondence with them.[35] With the notable exception of Newton's *Principia*, Petty never concealed his lack of interest in other men's books, maintaining that 'you can draw more knowledge and satisfaction from two hours of your own meditation than from 12

agents, lots 814, 815, 817, 818 in the Phillipps sale, formerly Bowood Vols. 13–18) are now BL Add. MSS 72859–64.

31 Lansdowne, *Petty-Southwell Correspondence* (n. 7), p. 118; the surviving papers, formerly Bowood Box D, are now BL Add. MSS 72900, 72901.

32 BL Add. MSS 72856–7 (Bowood Vols. 5 and 8); BL Lansdowne MS 1228, ff. 20–37.

33 Hunter, *Science and Society* (n. 6), pp. 94–6; K. T. Hoppen, *The Common Scientist in the Seventeenth Century* (London, 1970), esp. pp. 26–7, 84–7, 96, 103–4, 147–8, 202–3; Michael Hunter, *Establishing the New Science* (Woodbridge, 1989), pp. 185–222, 310–16.

34 Petty's general incoming letters are in BL Add. MS 72850 (Bowood Vol. 6); further copies of letters from Royal Society associates concerning the 'double bottom' are in BL Add. MSS 72894 and 72895 (Bowood Vols. 20 and 21); his letters from Southwell are in Add. MS 72851 (Bowood Vol. 7).

35 In addition to those cited in Hull, *Economic Writings* (n. 7), i, lvi, there are letters from Petty to Hartlib in the Osborn collection, Yale University Library, and copies of letters to Anglesey in Petty's letterbook, BL Add. MS 72858, fols. 72, 81.

hours endurance of other men's loquacity'.[36] So this failure to preserve their letters is perhaps an indication that he did not set great value on their correspondence either, except, as in the course of his shipping experiments, when it served to further a particular project of this own.

The archive includes small groups of administrative papers concerning both societies, supplementing the records of the institutions themselves. Several relating to the Royal Society date from the period 1673–74, when Petty's formidable organising abilities were harnessed to the task of collecting arrears and reversing the decline in membership.[37] Some of his draft contributions to the Society's programme are also present, including essays on cloth manufacture for the history of trades project,[38] his 1674 'Essay on duplicate proportion',[39] and of course his collection of papers relating to shipping and the 'double bottom', which was sponsored by the Society in its early stages. But the greater part of the last group survive in transcripts only, made in the early nineteenth century by an agent of the 2nd Marquess of Lansdowne. Shortly after this the originals were sold or given away. Randomly divided into two portions, they came into the hands of private collectors and are now in the Bodleian Library and the National Maritime Museum respectively.[40]

While one can expect an individual's drafts and retained copies of his letters and papers to be part of his archive, the letters and papers actually sent, where they survive, will usually be preserved elsewhere, and cannot properly be regarded as such. However in Petty's case a substantial portion of such material, particularly the important series of letters and papers sent to his cousin and most trusted agent John Petty in the late 1650s and early 1660s, and to his friend Sir Robert Southwell from the 1670s onwards are all now preserved as part of his archive. The former appear to have been returned after John Petty's death in 1671, though the sequence is not now complete.[41] The Southwell group is much larger. From the 1670s onwards, Southwell corresponded regularly with Petty, solicited information about his work and begged copies of any writings he mentioned. He then, as he

36 Fitzmaurice, *Life* (n. 7), p. 46: Petty to Boyle, 15 April 1653.

37 In BL Add. MS 72898 (Bowood Box H), fols. 17-32; see also Michael Hunter, *The Royal Society and its Fellows 1660–1700: the Morphology of an Early Scientific Institution*, 2nd edn (Oxford, 1994), pp. 93–7, and *Establishing the New Science* (n. 33), pp. 185–222, passim. The register books of the Royal Society, of which Sloane MS 243 is a copy, and the Dublin Philosophical Society, now BL Add. MS 4811, both contain records relating to him.

38 For Petty's papers on cloth manufacture and dyeing, see BL Add. MS 72897 (formerly in Bowood Box G), fols. 1–37; see also Keynes, *Bibliography* (n. 14), nos. 7 and 15, and Houghton, 'The History of Trades' (n. 9), pp. 49–52.

39 The manuscript of 'Duplicate Proportion' (formerly Bowood J9) is in BL Add. MS 72898, fols. 50–69; see also Keynes, *Bibliography* (n. 14), no. 16. For Petty's response to criticism of it, see BL Add. MS 72850, fols. 220–3.

40 Slatter, 'Report' (n. 7), pp. xi, 200–21. The Bodleian group is now MS Lyell Empt. 32 (the family papers of the later eighteenth century at Lyell Empt. 34–36 probably left the archive at the same time). The National Maritime Museum group (SPB/16), which has not been identified with Petty's archive hitherto, consists of 54 documents, 1662–85, some in Petty's hand. A label of James Matthews, librarian to the 1st Marquess of Lansdowne, establishes that it was once part of the archive.

41 BL Add. MS 72850, fols. 1–123, passim.

explained to Petty, systematically preserved everything he received in an ebony cabinet, 'as in an archive'; 'I shrine all up', he added, 'and fancy that in after times I shall be resorted to for your workes.'[42] About half of this collection appears to have been returned privately to Petty's family. Most of the rest was bought back by his descendants when the archive of Southwell's descendants, the De Cliffords, was sold in the 1830s.[43] Since Petty had also preserved many of the letters which Southwell wrote to him, the correspondence and most of its enclosed writings, though awkwardly bound in three sequences, is now substantially complete on both sides and a major source for the prolific last decade of Petty's life.

But by far the largest and most important component of Petty's archive, as least as large as all the others put together, is the collection of hundreds of his post-Restoration manuscript treatises and working papers. To judge from his own lists of his writings and projects, and from that compiled in the 1780s by James Matthews for the 1st Marquess of Lansdowne, this portion of the archive survives more intact than any other.[44] In fact, since the papers which Southwell, in his role as 'archivist', preserved independently have since been added, it is richer than Petty himself left it. Selections have been published, notably in the excellent edition by the 6th Marquess of Lansdowne entitled *The Petty Papers*, but these represent only a fraction of the total number. In fact the form in which this part of the archive has been preserved has made it difficult to use as a whole.

In 1685, just before he left Ireland for the last time, Petty recorded that he had 'rumaged and Methodized' his papers and stored them in fifty-three chests.[45] The nature of the methodising is not known, but it may well have followed the very broad categories of his 'Catalogus Scriptorum' of 1686: 'Designs of Profit', 'Engines', 'Workes' (i.e. extended treatises), '[writings or projects] about Ireland' and about 'Publiq Matters' in general.[46] But any effect of this methodising must quickly have been overlaid by the quantities added and removed after it took place. As we shall see, Petty's most prolific period of writing came after 1685, and after his death his papers were transferred from house to house by his sons, eventually being inherited in the mid-eighteenth century by the family of his only daughter.

42 Lansdowne, *Petty-Southwell Correspondence* (n. 7), pp. 34, 102, 113: 15 Sept. 1677, 11 Sept., 28 Nov. 1682.

43 The volumes which appear to have been given back without passing through the sale rooms are BL Add. MSS 72852–3 (Bowood Vols. 9, 10); those in the De Clifford sale are now Add. MSS 72854–5 (Bowood Vols. 11, 12). Excerpts from the sale catalogues are in Add. MS 72905; see also Slatter, 'Report' (n. 7), p. 98. Abraham Hill, FRS, made copies from some of the short essays enclosed in these letters, which are now BL Sloane MS 2903, fols. 13–54; it was also he who placed the Southwell manuscripts in order in 1703 and added the note that they 'doe highly merit to be preserved' (BL Add. MS 72854, fol. 1). Further authoritative texts of Petty's major writings were sold separately at the De Clifford sale and eventually acquired by the British Museum; they are now BL Add. MSS 21127–8. Most of the letters are printed in Lansdowne, *Petty-Southwell Correspondence* (n. 7).

44 For several of Petty's lists, see Lansdowne, *Petty Papers*, ii, 259–69; for Matthews' list, see Slatter, 'Report' (n. 7), p. iv. The groups on coinage and banking listed by Matthews are now unlocated. The letters to Henry Cromwell, acquired by Petty as his secretary, which are the last items on Matthews' list, are now BL Lansdowne MSS 821–3.

45 Lansdowne, *Petty-Southwell Correspondence* (n. 7), p. 138: 23 April 1685.

46 Lansdowne, *Petty Papers* (n. 7), ii, 264–6.

The most like to him of all his children, she had brought by her marriage into the Fitzmaurice family, according to her grandson the 1st Marquess of Lansdowne, 'whatever degree of sense may have appeared in it, or whatever wealth is likely to remain in it'.[47] Over this period, as we have seen, the archive suffered further losses and disorder. The first attempt to remedy them was made by James Matthews, librarian to the 1st Marquess of Lansdowne in the 1780s. He arranged the remaining papers into subject bundles, under such headings as 'Political, Irish'; 'Political, English'; 'Medical Studies'; 'Royal Society'; 'Military Projects'; 'Woollens'; 'Money & Coinage'; 'Trade'; 'Forests'; 'Religion'; 'Reform of Church Temporalities'; 'Family Occurrences': and so on. Although the order of the groups has been changed, they have remained the basis of the arrangement of the papers ever since. But no attempt was made, then or later, to establish chronological or textual order within these subject groups.[48]

In the 1970s a detailed description of the papers in this form was compiled by a professional archivist, Miss M.D. Slatter. Although it is admirably thorough and accurate, in the absence of systematic order amongst the loose papers it can be only a piece by piece list of each item. It records their publication in modern editions, but in most cases not their relationship to one another and only a minority of the many undated items have been assigned dates. The disadvantage of this method in the case of Petty is that a good many of his papers are very difficult to describe in isolation, since they consist of sequences of numbered short paragraphs, sometimes untitled and often on a variety of overlapping topics.[49] This is not a great problem when subject groups are small and specific, 'Woollens' or 'Military Projects' for example. But it makes the largest categories such as 'Political, Irish' and 'Religion' (much of which also refers to Ireland), consisting of hundreds of individual papers in no archival order, extremely difficult to assess systematically. It is significant that even those with the most detailed knowledge of the archive have dealt only cursorily with these categories, even though they are by far its largest component.[50]

Now that the archive has a permanent home in a public repository, something can and should be done to improve this situation. It seems sensible to retain the broad subject groupings, which have a long history; although they do tend to obscure both the great concentration of writings in Petty's last years and the way in which many of his projects, his 'engines', his commercial proposals and his social and political reforms, interlocked into one large scheme at the end of his life. But the first question to ask of a document in an archival context is not what is it about, but in what historical process, institutional, personal or intellectual, did it originate. It is important that documents be presented in a way which reflects their chronological and textual order.

47 Slatter, 'Report' (n. 7), p. v; Fitzmaurice, *Life* (n. 7), p. 312.

48 Slatter, 'Report' (n. 7), pp. iii–iv; see also the comment by the 6th Marquess of Lansdowne, *Petty Papers* (n. 7), i, xx: 'the arrangement . . . has never been seriously attempted'.

49 Slatter, 'Report' (n. 7), p. vi.

50 From the selection printed in Lansdowne, *Petty Papers* (n. 7) it is not apparent that these papers form the largest portion of the archive; Sharp, 'Sir William Petty' (n. 8), does not discuss them in detail.

There is plenty of evidence from which to do this. Petty himself dated about a third of his papers with the year of their composition, and many of the others can readily be dated at least as closely, on the basis of changes in his handwriting (partly as a result of his deteriorating eyesight), and that of his succession of amanusenses, his dated lists of his writings, references in his correspondence and in the dated works themselves. The establishing of textual relationships is less straightforward. Because, as we shall see, Petty was reluctant to revise his work, there is not invariably a clear progression through various drafts to a finished version. Nevertheless a good deal can be done to reduce the confusion of unidentified and separated drafts, synopses, versions and copies of the same or closely related items.

A number of specific discoveries and general conclusions have already emerged from this process of restoring some kind of archival order to Petty's working papers. Of the former I will mention only two examples here: the bringing together of previously separated and misidentified early drafts of his most famous tract 'Political Arithmetick';[51] and the identification of autograph notes by the Marquess of Halifax relating to Petty's last work, 'A Treatise of Ireland', and their incorporation by Petty into his final text: one of several scraps of evidence from the archive of the contact between the two at this period.[52]

But two general points which have clearly emerged deserve more emphasis. Those who have studied Petty's ideas have emphasised the importance of the pre-Restoration period for his intellectual development.[53] But the fact remains that his archive contains far more writings of the second half of his life than of the first, and that this is not, as might seem likely, just an accident of survival. There were certainly some early projects of which Petty himself did not keep a record, and it is fortunate that Hartlib in some cases did this for him;[54] though this suggests that they might have been more important to Hartlib than they were to Petty. But there is no doubt that Petty's output did increase dramatically in his later years. In 1671 he compiled a year by year list of his writings, clearly intended to be all-inclusive, beginning from the time he first put pen to paper in his teens. It contains about fifty items in all.[55] For the rest of Charles II's reign, the years 1672–84, about eighty-five more can now be identified.[56] But for the period 1685–87, the last three years of Petty's life, the process of dating and sorting which I have mentioned identifies more than three hundred separate writings.[57]

51 BL Add. MS 72865, items 4 and 5.

52 BL Add. MS 72886, items 4 and 5. Petty dedicated his 'Quantulumcunque concerning Money' to Halifax in 1682; Petty is not mentioned in *The Works of George Savile, Marquess of Halifax*, ed. Mark N. Brown (3 vols., Oxford, 1989).

53 E. Strauss, *Sir William Petty* (London, 1954), pp. 138–41; Sharp, 'Sir William Petty' (n. 8), p. x; T. Aspromourgos, 'The Life of William Petty in relation to his Economics', *History of Political Economy*, 20 (1988), 28–45, on p. 38.

54 Sharp, 'Sir William Petty' (n. 8), pp. x, 82. The Hartlib Papers in Sheffield University Library include material by Petty on dyeing, weaving and various mechanical inventions: see the CD-Rom edition by UMI (Ann Arbor, 1995).

55 Lansdowne, *Petty Papers* (n. 7), ii, 260–2.

56 BL Add. MSS 72865, 72878–80.

57 BL Add. MSS 72866–7, 72881–9.

One further important point: of the three hundred or more papers of James II's reign, more than two-thirds are about Ireland and many more touch in some way on Anglo-Irish relations. In fact this balance holds true for the whole of Petty's output. Of the fifty-nine volumes in which the archive is now arranged, thirty-five are entirely about Irish matters, and many more, including the correspondence, contain related material. The only substantial part of the archive which has no relevance at all to Ireland is the collection of early medical manuscripts, since these precede Petty's first crossing to Dublin in 1652. Even the 'double bottom' experiments were conducted largely from Irish shipyards. The conclusion we must draw is that it was not the intelligence networks of Samuel Hartlib or the London bills of mortality or the gatherings of the Royal Society which were Petty's chief sources of intellectual stimulus; it was Ireland. Much of this is not easy material to deal with. It has not been its disordered state alone which has caused Petty's admirers to leave it to one side. His projected social and economic 'reforms' for Ireland, culminating in a plan for a mass transplantation of its population, were more radical than any he dared propose for England, and it was clearly a large part of Ireland's attraction for him that it appeared to offer scope for this kind of experiment. It is a striking reminder that, in Roy Foster's words, 'the late seventeenth-century approach to learning sees Ireland as a laboratory';[58] and it brings us back to Petty's affirmation, 'the seat of project and designe must be there at last'.

It remains to examine in more detail the form which these writings of Petty's last years take and what ends he had in producing them. In his own lists he did not distinguish between his published and unpublished works, and in fact those printed in his own lifetime were very few as a proportion of the whole. The handful before the Restoration were produced for specific promotional or polemic purposes; and having published the *Treatise on Taxes* in 1662, Petty refused either to acknowledge authorship or have it reprinted afterwards. From then on, whenever he wanted one of his works to be more widely known, he preferred circulation in manuscript or venues such as the Royal Society. In the late 1680s he did again venture into print with his essays on the populations of London, Paris and Rome, but presumably because this was the best way of bringing them to the attention of foreign scholars.[59] In general the manuscripts of those few works actually sent to the printers by Petty do not survive in his archive.

A preference for coterie circulation over print publication was of course common enough at this period. In fact the term publication can legitimately be applied to scribal as well as printed circulation in the seventeenth century.[60] But for Petty it was yet another manifestation of his sense of being, as he put it, 'in a

58 Roy Foster, *Modern Ireland 1660–1972* (London, 1988), p. 118. On this point, see also Terence Ranger, 'Strafford in Ireland: a Revaluation', in Trevor Aston (ed.), *Crisis in Europe 1560–1660* (London, 1965), pp. 271–94, on p. 281.

59 Keynes, *Bibliography* (n. 14), pp. 1–29; *Two Essays in Political Arithmetick concerning the People, Housing Hospitals, &c. of London and Paris* (1687) was first published in French; see Hull, *Economic Writings* (n. 7), ii, 502.

60 On this subject, see Harold Love, *Scribal Publication in Seventeenth-Century England* (Oxford, 1993).

captious world from which I use to find little favour'. The experience of conducting his shipping experiments in public, and in the process enduring 'the impertinencies, derisions &c of 99 in the 100 of the whole world', confirmed his preference for confining his studies to his own and other men's closets. His ideal was to have copies of his works 'in every great mans Archives', where they would have most chance of being influential, without being exposed to general public comment and criticism. Given the controversial content of much of his political and economic writing, this was also safer. 'You know I have no luck with my politicks', he wrote to Southwell in refusing to have his tract 'Political Arithmetick' printed in the 1670s, 'Slight court tricks have advanced many men, but the solid study of other men's peace and plenty ruines mee.'[61] Thus, although he wrote sheaves of detailed replies to published attacks on the Irish land settlement in 1685 and 1686, he kept the manuscripts by him pending an official sanction to publish which never came, and they still remain as he left them in his archive. The circulation list attached to one of them includes the present and past Lords Lieutenants of Ireland, Ormonde and Clarendon, the Lord Treasurer and the Lord President, the King's advisers, Dartmouth, Halifax and Tyrconnel, and the government censor Sir Roger L'Estrange, as well as friendly associates such as Pepys, Aubrey, and Robert Boyle's sister, Lady Ranelagh, a leading representative of the Irish landowners in London. But, 'Pray Madam', Petty wrote nervously to the last, 'shew them onely to the true friends of the King, Church of England & Emprovement of Ireland, and if it bee possible secure mee against suffering for those 3 Interests.'[62]

Drafts or early versions of the best known of Petty's economic tracts which were in manuscript circulation in his lifetime are present in his archive; these include 'Political Arithmetick', 'The Political Anatomy of Ireland' and the 'Quantulumcunque concerning Money'. But the majority of the writings preserved there are much shorter and less finished, consisting usually of a bifolium or at most a few sheets of numbered points or short paragraphs. Aubrey, seeing these in Petty's archive shortly after his death, called them 'tractatiuncli'.[63] A few are synopses of large-scale projected works, intended for eventual publication but never completed; examples are a philosophical work entitled 'The Scale of Creatures', 'A Treatise of Naval Philosophy in three Parts', and 'A Dictionary of Sensible Words'.

But most are shorter and more specific: notes on an apparent miscellany of socio-economic projects and practical inventions (what Petty himself called 'engines'): for water pumps, for making seawater fresh, for war chariots and land carriages, for academies of science and technology, for reorganising the post, for replanting royal parks with timber and hemp for the Navy, for dealing with bankrupts, for providing work for the poor, or (most weirdly) for the use of the

61 Lansdowne, *Petty Papers* (n. 7), ii, 171, and *Double Bottom* (n. 7), p. 35; Petty to Dr Robert Wood, 10 April 1675, BL Add. MS 72858, fol. 211; *Petty-Southwell Correspondence* (n. 7), pp. 61, 55; Fitzmaurice, *Life* (n. 7), pp. 201–2.

62 Drafts of 'Speculum Hiberniae', 1686, BL Add. MS 72884; for Lady Ranelagh, see Sir Alan Broderick to Ormonde, 14, 31 Dec. 1667, Bodleian MS Carte 36, fols. 31, 63.

63 Aubrey, *Brief Lives* (n. 4), ii, 146.

glut of coarse Irish wool in the construction of fortifications and harbour walls. There are several London projects: for a wall around the city, a plague hospital, a new bridge at Lambeth. There is a further group on a colony in Pennsylvania. But by far the largest collection relates to matters of government: schemes for increasing and registering the population and for the reform of taxation, the law courts, the electoral system, the church revenues, the King's councils, and above all Ireland: its land settlement, religious organisation and in the end its entire population. Many of these papers do not appear to have been intended even for private circulation, and there must have been other reasons for their production in such numbers.

Petty had always used 'ratiocination' as a form of mental exercise and recreation. Some of his comments suggest that he wrote, even on public matters, almost as a form of mental hygiene: 'to rid my head of so many troublesome conceits', 'to vent the superfluous humour fermenting within myself', or as a relief from the stresses of his lawsuits.[64] As his active life dwindled and the frustrations of his public ambitions multiplied, he turned increasingly to this occupation for consolation. 'You will ask me why I persist in these fruitless Labours', he wrote to Southwell, 'I say they are Labours of pleasure, of which ratiocination is the greatest and most Angelicall . . . Contrary to the nature of Sentiall pleasures, It continueth to extream old age and shall goe with us to heaven; and I had rather represent heaven by the facultyes and exercises of sublime Ratiocination Than (as some doe) by Musick, or than (as others doe) by beauties.'[65]

It was always the expounding of a concept in broad terms which most appealed to him, not the labour of expanding, 'polishing' or buttressing it with evidence, and this accounts for the unfinished nature of most of his manuscripts. Although he certainly needed statistics as raw material for his works on economics, taxation and demography, he did not have much personal enthusiasm for 'all this laborious bustling and groping', as he termed empirical methods of accumulating data,[66] and in any case much of what he needed was simply not available. When the errors and inadequacies in his calculations were pointed out to him, he only remarked that this was according to his 'usuall way in grosse draughts of new conceptions'.[67] Of one of his early exercises in political arithmetic, consisting of a series of very miscellaneous numbered jottings, he remarked disarmingly, 'Some of them are not so very true as I could have made them . . . Some I knew not how to get downe truer; but they are all true enough to prove what is intended.' This method, he explained further, was intended 'to prevent the labor of writing long books'.[68] When his publisher complained that one of his demographic tracts was not of

64 'Treatise on Taxes', in Hull, *Economic Writings*, i, 5; 'Verbum sapienti', ibid., i, 119; Lansdowne, *Double Bottom* (n. 7), p. 107; Lansdowne, *Petty-Southwell Correspondence* (n. 7), p. 118: 25 Feb. 1684; see also the excellent discussion by the 6th Marquess of Lansdowne in *Petty Papers* (n. 7), i, xxi–xxv.

65 Lansdowne, *Petty-Southwell Correspondence* (n. 7), pp. 283, 295–6.

66 Hull, *Economic Writings* (n. 7), ii, 394.

67 Ibid., i, lxviii; Petty to Wood, 10 April 1675, BL Add. MS 72858, fol. 211.

68 Lansdowne, *Petty Papers* (n. 7), ii, 226–7.

sufficient bulk, Petty only replied that he wished all books were of less bulk.[69] In fact once he had put down the broad outlines of an idea in manuscript they had the same weight in his mind (and in those of his admirers) as a completed and published work. He claimed for example that a Dutch book on naval matters which was 'infinitely cry'd up at Court' (probably Nicolaas Witsen's *Book of Building Ships* of 1671) did not contain a fraction of what was in his own 'Treatise of Naval Philosophy', although the latter, as is evident from the copies in his archive and elsewhere, existed only in a few pages of synopsis passed around in manuscript amongst his friends, who hoped that one day he might be persuaded to expand it.[70] The most striking example of Petty's preference for this kind of format is his work on the reformation of the church revenues in England and Ireland in 1686. Instead of presenting his project connectedly in a single treatise, he produced no fewer than thirty short papers on specific aspects, each complete within itself and some present in more than one draft or version.[71]

Yet the motives for Petty's increased output in his last years were certainly not as purely personal or 'Angelicall' as some of his comments suggest. By this time he was an old man in a hurry. Producing a treatise in any form still interested him much less than the public adoption of his schemes and the realisation of his ambitions. Whereas Charles II and the Duke of Ormonde had disregarded him, James II and Tyrconnel were prepared to grant him interviews and read some of the papers he submitted, provided at least that he kept them short. In their readiness to make radical changes to the government and religious establishment in England and Ireland, Petty evidently glimpsed something of the opportunities which the Interregnum had opened up to him, and he trimmed his sails accordingly. 'I am beginning the world again', he wrote excitedly to Southwell, 'and endeavor instead of quarelling with the King's Power, to make him exert all hee hath for the use of his subjects.'[72]

One of his drafts of this period concerning the best means of promulgating his schemes does much to explain the form many of his papers take:

> What should a private man doe, to whom God hath given some extraordinary talent for the publick good. He should if he hath meanes of subsisting make the Improvement of that Talent his pastime and Recreation 2. Entertaine his Ingenious friends with the produce of his studyes 3. make the same known to the King and to be understood by himself personally 4. Instruct all persons in the same who have easy access unto, or credit with the King . . . 5. he should adapt his proposals for universall good, to the Extraordinary and eminent advantage of particular favorites.[73]

69 Quoted in Strauss, *Petty* (n. 53), p. 179.

70 Petty to Wood, 30 Dec. 1672, McGill University Library, Osler MS 7612, fol. 49. Contemporary manuscripts are National Maritime Museum, SPB/16, No. 52, and BL Add. MS 72854, fols. 99–104 (Southwell's copy). Cf. Wood to Pepys, 17 June 1682, Bodleian MS Rawl. A. 178, fol. 153; Pepys to Wood, 16 June 1682, Rawl. A. 194, fol. 279. For its eventual publication in 1691, see Keynes, *Bibliography* (n. 14), no. 39.

71 BL Add. MS 72889 (from Bowood Box I).

72 Lansdowne, *Petty-Southwell Correspondence* (n. 7), p. 240.

73 'A Dialogue between C: and D:', BL Add. MS 72866 (formerly Bowood Box B 81), fols. 146–7.

From this time onwards, his archive shows him aiming his schemes directly at the King, his ministers, advisers and favourites. Hence the increasing numbers of papers after 1685 listing or briefly expounding his projects and ideas, with such titles as 'Intimations to the King', 'Remedies for the King', 'Advantages humbly offered to the King', 'Things which may be done by the Kings Authority and prerogative', 'Things which the King may probably obtaine from the next Parliament', 'Papers delivered to the King', or simply 'Transactions at Windsor 19th of September 1686'. Others of 1686 are endorsed 'A Memento to the Lord Tyrconnel' or 'The Scope and Designe of the Papers delivered to the Earle of Tirconnell'.[74] In his list of 'undertakers' of 1687 for the commercial projects of a new bridge, a national bank and the use of royal forests to supply timber for the Navy, the prime movers are set down as himself and his Royal Society associates Sir Peter Pett and Sir James Hayes, together with Sir Anthony Deane, Samuel Pepys and Wren as naval and architectural experts. But others invited to offer their support in return for profit include the Lord Chancellor, the Lord Treasurer, Lords Sunderland, Dartmouth and Halifax, and two of the King's leading courtiers, his Groom of the Stole, Lord Peterborough, and his favourite, Lord Churchill, the future Duke of Marlborough.[75]

In his papers on these subjects Petty was clearly doing what might be termed making a pitch: attempting to grab the attention of those in positions of power and influence by presenting them with striking, profitable, easily digested proposals, with the promise of being able to expand and demonstrate any of them further if invited.[76] He has sometimes been represented at this time as a somewhat pathetic and deluded figure, heaping up his offerings of unsolicited and unheeded advice to the great. But the opinion of James II's most authoritative biographer John Miller, based on the limited number of papers which are in print, is that Petty's efforts deserve to be considered seriously.[77] In fact this whole body of material calls for further attention, not just for what it can reveal of Petty's or of the King's mind, but for what it has to say of the possibilities inherent in this short but radically destabilising reign.

But of course it was all much too late. By this time Petty had only a few months to live and preparations were already afoot amongst the King's Protestant subjects which would drive him out of his kingdom a year after that. Southwell told Petty plainly what he must have known very well himself by this time, that he could never hope to see such complex projects implemented in his lifetime. But he consoled Petty that future generations could still benefit from his ideas and promised to help preserve his writings and prepare them for the press. But even Southwell, 'that much enduring man', as one of Petty's editors has called him,

74 BL Add. MSS 72866 and 72885, passim; a 'Breveat of the Church Temporalities' and 'An estimat of what may be gayned by the Church reformation in Ireland', BL Add. MS 72889, fols. 24–7v, 30, are also endorsed by Petty as given to the King, 29 Oct and 10 Nov. 1686.

75 BL Add. MS 72893 (formerly Bowood Box G, 64), fol. 67.

76 Lansdowne, *Petty-Southwell Correspondence* (n. 7), p. 282.

77 John Miller, *James II: a Study in Kingship* (Hove, 1978), pp. 167–8.

became alarmed at the sheer numbers of papers which were mounting up in Petty's closet and felt that he was taking too little care to leave them in a publishable form. He meaningfully cited the example of Dr Isaac Barrow, who left only a handful of papers to his executor, 'soe carefull had he beene to finish all he ever tooke in hand, either printing what he wrote, or leaving his thoughts and collections all ready for the Presse'.[78]

'As to my papers', Petty replied defensively, 'Those relating to Ireland (which are neare five hours reading) are corrected ready for any use, and soe are a bundle of others concerning particular designs. The rest I will finish as I can and as they deserve . . . I have many Important papers redy for the presse, and many more Intelligible and usefull perhaps not fitt to bee printed. Others I perfect daily.'[79] The results of this last minute attempt to 'perfect' can clearly be seen in his archive. A number of short tracts have been fair-copied and sewn in gatherings. One longer tract of 1687, entitled 'Magnalia Regni', attempts an amalgamation and detailed exposition of many of his individual projects for reform, from a complete survey of the lands and people and a union between England and Ireland, to a new modelling of the armed forces, the reform of Chancery, the establishment of a land registry, a national bank, a plague hospital and academies of mathematics and mechanics, and the improvement of highways and the post. Another of the same year, 'A Treatise of Ireland', elaborates his proposals for a mass transplantation of the Irish population.[80] But it is clear that these lengthy expositions were forced and uncongenial labour for Petty. The former peters out into a characteristic synopsis half-way through, and the latter has no less than three different title pages and a complicated and wishful table of contents with so little clear relationship to the main text that its modern editor felt obliged to provide his own concordance between the two.[81]

Southwell did assist in publishing authorised versions of Petty's main economic treatises in the 1690s, though not without some difficulty from Petty's widow, who considered herself bound by his reluctance to publish in his own lifetime.[82] But the bulk of his papers proved too large, too disorganised and too controversial for publication in the Williamite era and remained in seclusion in his archive. Now that this has a permanent home in a public institution and the order of the working papers more closely reflects the process of their composition, the way is open for a fuller assessment of Petty's many-faceted contribution to the Scientific Revolution, and particularly of the central role it gave to Ireland, than has been possible before now.

78 Lansdowne, *Petty-Southwell Correspondence* (n. 7), pp. 281, 287, 289.
79 Ibid., pp. 293, 299.
80 In BL Add. MS 72866, fols. 118–29v (formerly Bowood Box B, 108); Add. MS 72886 (Bowood Vol. 4).
81 Hull, *Economic Writings* (n. 7), ii, 545–53.
82 Ibid., i, 238; Keynes, *Bibliography* (n. 14), p. 34.

Summary List of the Petty Papers (Add. MSS 72850–72908), with former Bowood Numbers

Correspondence

72850: General correspondence (Vol. 6)
72851–5: Southwell correspondence (Vols. 7, 9–12)
72856–7: Family correspondence (Vols. 5, 8)
72858–64: Letterbooks (Vol. 19, 13, 16, 14, 15, 17, 18)

General Economic and Political Affairs

72865: papers, *temp.* Charles II (Boxes A, B, C, E, G)
72866: papers, *temp.* James II (Boxes A, B, C, E)
72867: papers concerning London, Paris, New England (Box E)

Ireland

72868–73: Down Survey of Ireland (Vols. 24–8)
72874: History of the Down Survey (Vol. 23)
72875: maps of the *Hiberniae Delineatio*
72876: 'Census'; c. 1659 (Box 12)
72877: papers, 1653–59 (Box F)
72878: papers, 1660–72 (Box F)
72879: papers, 1670s (Boxes B, C, F, Vol. 1)
72880: papers, 1680–83 (Box C)
72881: papers 1684–85 (Box B)
72882–5: papers, 1685–86 (Boxes B, C, Vols. 1, 4)
72886: drafts of 'A Treatise of Ireland', 1687 (Box B, Vol. 4)
72887–9: papers on religious issues (Box I)

Medicine, Science, etc.

72890: Statistical papers (Box A)
72891–2: Medical papers (Vol. 3, Box H)
72893: Papers relating to shipping, the admiralty, etc. (Box G)
72894–5: 'Double Bottom' transcripts (Vols. 20–1)
72896: Anonymous treatise on navigation (Vol. 22)
72897: Miscellaneous schemes and inventions (Boxes G, H, E)
72898: Papers relating to the Royal Society, etc. (Box H)
72899: Verse (Vol. 2)

Kerry Estate Papers

72900–2: seventeenth-early eighteenth century (Boxes D, F, 14)
72903–4: eighteenth-nineteenth century (Boxes 14–15)

Papers relating to the Archive

72905–7: Notes and transcripts, eighteenth-twentieth century (Boxes J, K)
72908: Lord Kerry's Life of Petty [1830s] (Box 17).

6

The Archives of Christiaan Huygens and his Editors

Joella G. Yoder

Although Christiaan Huygens willed his manuscripts directly to a library, his papers have not come down to us in the condition in which he left them. His editors have strongly influenced the current state of the collection, most particularly with regard to the ordering of the unbound papers. Hence my title is deliberately ambiguous, for I would claim that, in fact, the Codices Hugeniorum are the archives of both Huygens and his editors.[1]

Christiaan Huygens shocked his family by leaving his worldly goods to people other than those who were expected to inherit his estate. Among those non-traditional bequests was one to posterity, for he left his papers to a public institution. The act shows that he had a very strong sense of history and of his own importance. He specified in his will that his 'writings on Mathematics', consisting of nine bound books lettered *A* to *I* and many additional treatises, go the library of the university at Leiden. He further requested that professors Burchard de Volder of Leiden and Bernard Fullenius of Franeker edit those pieces that they felt ready for publication.[2] Huygens specified four works for consideration: (1) his treatise on dioptrics, (2) his study of perihelia, (3) his 'laws of percussion in colliding bodies etc', and (4) his instructions in Dutch for grinding lenses. In order to facilitate their task, the will permitted de Volder and Fullenius to take home the pertinent manuscripts, with the obvious stipulation that the papers be turned over to Leiden when the editing was completed. Leiden was also to receive his 'letters from professors and other learned men' as well as the 'French letters' from G.W. Leibniz and the Marquis de l'Hospital lying 'in a separate drawer along with my answers'.[3]

1 I have written on this subject before; see 'Christiaan Huygens' Great Treasure', *Tractrix*, 3 (1991), 1–13. I have tried to make the current article supplement the earlier one and to avoid repetition as much as possible. For example, that article describes more thoroughly the current arrangement of the augmented Codices Hugeniorum. I am preparing a catalogue of the collection that cross-lists the manuscripts with the *Oeuvres complètes*.

2 Both men were well qualified to appreciate Huygens' work. Fullenius and Huygens had corresponded over problems in dioptrics during the 1680s. Among other ties, de Volder was one of the referees for Huygens' report to the East Indies Company on the use of his clocks to find longitude at sea. The two editors were to receive 1000 guilders each for their effort.

3 'Ick legateere mede aen deselve Biblioteecq, de pacquetten daerop geschreeven staet Literae Doctorum off Eruditorum . . . alsmede de Fransche brieven van Monsr. Leibnitz en den Marquis de l'Hospital,

What did Huygens intend by this bequest? The four works that he specified for publication were, in fact, in a condition that required very little editing, since all four existed in reasonably clean drafts or copies. *De Motu Corporum ex Percussione* and *Dioptrica* could have been sent as is to the printer by Huygens himself. The instructions for grinding lenses were in Dutch and, although Huygens probably intended that they be published in the vernacular for the sake of the artisans, Fullenius and de Volder translated the instructions into Latin under the title *Commentarii de Formandis poliendisque vitris ad Telescopia*. Likewise, the draft of the treatise on coronas and false suns, which was partially in Dutch, was completely translated as *Dissertatio de Coronis et Parheliis*. Although the manuscripts thus required some treatment before they could go to press, Fullenius and de Volder clearly did not need access to the complete corpus in order to carry out their commission. Huygens could have left only those four specific items to the care of his chosen editors, just as he did with the draft of *Cosmotheros*, which was already in press and which he 'recommended' in his will to his brother, Constantijn, to whom it was dedicated, for the completion of its publication.

That being the case, why did Huygens give the entire set of his workbooks and papers to the academic community? Perhaps the manuscripts were to play witness for any claims of priority against his inventions. Perhaps he remembered how poorly he and his brothers had served their father's request that they arrange *his* papers for publication and he wished to have someone outside the family tend to his intellectual legacy. Indeed, we know from notes by David Gregory that brother Constantijn opposed the publication of the instructions for grinding lenses, which he had cowritten with Christiaan.[4] Thus, its appearance on the list of four is particularly curious. Ironically, one of the family members did tend to Christiaan's intellectual legacy. His brother Lodewijk's eldest son, Constantijn Lz, the nephew who most benefitted from the non-traditional will, preserved those manuscripts that had not gone to the University of Leiden and actively assisted in a later edition of Christiaan's works made by the Newtonian physicist Willem's Gravesande.[5]

Whatever the motivations for the bequest, the man who published relatively little in his lifetime relented and left us not only treatises ready for publication but also the raw material for many more. The bounty was too tempting for de Volder and Fullenius. They interpreted their mandate broadly and edited, in a much more active way, two additional manuscripts, *De Vi Centrifuga* and *Descriptio Automati*

leggende int groote Sakerdane Cabinet tot Hoffwijck in eene lae[t]ije apart, waerbij sijn mijne antwoorden' [*Oeuvres complètes des Christiaan Huygens* vol. 22, 775–7, on p. 776].

4 '26 May. 1694. Londini. Monsr. Hugens offered to publish his excellent way of grinding glasses but was diverted by his brother of Zulichem' [Royal Society MS 247, fol. 76]. Also, 'May 1703. M. Hugens book [the de Volder and Fullenius edition, one presumes] much commended. The manner of grinding Glasses true & honest his brother resisted the publishing of it all his time' [MS 247, fol. 63]. Constantijn Jr died in 1697.

5 Constantijn Lz (also known as Constantijn IV) made copies of letters that were in his possession and loaned them to 's Gravesande. At least three of the books of copies are extant: two in Hug. 36 and one in the Koninklijke Bibliotheek (The Hague), KA XLV. Hug. 32 contains evidence of his activities, including some letters to 's Gravesande and a sketchy biography of Christiaan.

Planetarii,[6] thus initiating a sequence of editorial intrusions that significantly increased the number of published works by Huygens and at the same time significantly obscured the manuscripts from which they were drawn.

The manuscripts bequeathed by Huygens were first catalogued in 1716, as part of a general survey of the collections at Leiden.[7] I had long presumed that the 1716 catalogue does not reflect his original arrangement of the papers, because five of the six pieces edited by de Volder and Fullenius are listed as being in one codex, a highly suspect coincidence given that they concern very different topics. However, when I undertook to reconstruct the original order, I discovered a more complex, and ultimately frustrating, story.

The 1716 catalogue lists eighteen bound books with summaries of their contents. The order of the books has remained intact to the present day (Hug. 1–18), including the oddity that workbooks *A* to *I* were not, and consequently still are not, numbered in alphabetical and thus chronological sequence. Nor does the catalogue even refer to them by letter; only the dates given in their descriptions pinpoint them.

The catalogue then assigns a separate numerical sequence to thirteen bundles of loose papers (fasciculi), which are also briefly described. Six bundles (fasc. 6–12) contained letters written *to* Huygens and are now gathered together in one massive codex (Hug. 45). Most of his drafts for replies to those letters, all family letters, and state letters, such as those pertaining to his appointment to the court of Louis XIV, were not originally transferred to the library. I do not know whether this sequestering occurred because Huygens intentionally phrased his will to effect such a division or, the more likely case, because his family chose to read the will narrowly. Not mentioned in the will, and thus not considered part of the bequest, were his many instruments, such as clocks, telescopes, and lenses. Fortunately, much of this material rejoined the papers in later acquisitions.[8]

The remaining bundles (fasc. 1–5 and 13), as described by the catalogue, contained Huygens' working papers, everything from whole treatises and drafts to scrap calculations and even the work of other researchers. Amazingly, the 1716 inventory of the first five bundles corresponds almost exactly to three lists Huygens himself made. His lists, in turn, are compilations of individual title pages still extant, such as 'Musica' (Hug. 27, fol. 5). This remarkable concurrence suggests that Huygens himself arranged his papers into small groups, such as a draft or a packet of related material, each with its own separate title page, and then collected the small groups together into three large sets, each ordered according to a list of multiple titles that probably acted as a cover for the whole set. This collection then passed essentially undisturbed to the library at Leiden, except that

6 The six works appeared under the title *Opuscula postuma* (Leiden, 1703).

7 'Manuscripta Praecipue Latina, ab Illustrissimo Christiano Hugenio, Zelemi Toparcha, Academie legata', pp. 351–7 in *Catalogus Librorum tam impressorum quam manuscriptorum bibliothecae publicae Universitatis Lugduno-Batavae* (Leiden, 1716).

8 A detailed history of the fate of his lenses is given by Anne C. van Helden and Rob H. van Gent in *The Huygens Collection* (Leiden, 1995). In particular, they recount Constantijn Lz's role in their preservation and sale.

the very large third set was subdivided, in part by Huygens himself. Subsequently, according to my reconstruction, the cataloguer repeated the covering lists when creating an inventory of the collection.

Most items on Huygens' three lists do correspond to a single title page [see Tables 1–3]. In many cases he either repeated the single title verbatim, foreshortened it to its beginning phrase, or made minor alterations in the wording. The accord is so close that, in a title that refers to youthful work sent to Mersenne, a crossed-out line regarding 'imperfect' algebra is repeated (with slight variation) and likewise crossed-out in the corresponding list. Indeed, only a garbled reference to a preliminary copy of the *Traité de la lumière* abandons the original wording entirely. Fourteen of Huygens' title pages are now duplicated, usually verbatim, within the manuscript collection by more modern covers, probably made by the library during nineteenth century cataloguing efforts. Curiously, four more library covers have no parallel single title pages in Huygens' hand, although they do match titles in one of his lists.[9] Have his title pages gone missing? Three of the four library covers concern divisions Huygens made in his papers on motion, and the only one of these that does not repeat exactly the title as worded on his list is the first, which leaves off the name of the amanuensis who made the clean copy of *De Motu* used by Fullenius and de Volder. It was Denis Papin, whose departure for London in 1675 thus limits the date of the copy.[10]

A detailed comparison of Huygens' three lists with the entries in the 1716 catalogue reveals a similar agreement between titles, with many congruences but some small differences. The cataloguer translated French and Dutch titles into Latin; for example, Huygens' idea for a method of printing, titled 'Imprimerie', became 'Nova chalcographiae ratio'. In the case of 'Pieces touchant les Regles de la percussion' the translation to 'Regulae percussionis' has the unintended effect of masking why those papers should stand apart under a French title, but the segregation at least remained. Naturally the cataloguer changed all first person references to the third person. Aside from these relatively minor changes, however, the description of fasc. 1 almost exactly repeats Huygens' phrasing in one list [see Table 1], except that the library catalogue says nothing about insignificant mathematics [item 3] or about the *Dioptrica* copied by Niquet [item 6], although the latter could be considered part of the previous title, 'working papers pertaining to Dioptrics'. Likewise, the description of fasc. 2 matches a second cover by Huygens [see Table 2], except, I presume, that the catalogue consolidates all titles involving 'adversaria' of *De Motu* while giving more details regarding the items sent to the Académie des Sciences. Huygens' third list has been split into three codices, fasc. 3, 4, and 5 [see Table 3]. The beginning of his list reappears, albeit in a different

9 Then there is the really odd, unmatched library cover: 'Chartae ad Cosmotheron pertinentes' [Hug. 28, fol. 121]. This title appears nowhere on Huygens' lists.

10 Somehow the editors of *Oeuvres complètes* lost track of this information: 'Nous ignorons à quelle époque le Manuscrit définitif . . . fut composé' [vol. 16, 14]. Not only do they not name Papin in the introduction to their edition of *De Motu* in vol. 16, but at one point they conjectured that the amanuensis was Fatio de Dullier; see their n. 3 as well as their internal notes on Hug. 26A, fol. 89.

Table 1

Single Title in Cod. Hug.	Multiple Title List Hug. 32A, fol. 15	1716 Catalogue fasc. 1
Hug. 25, fol. 147	De Problemate Alhazeni de puncto Reflexionis in speculo sphaerico. Constructiones Slusij et nostrae, cum literis Oldenburgij.	De problemate Alhazeni de puncto reflexionis in speculo sphaerico. Constructiones Slusii, & Hugenii, cum literis Oldenburgii.
Hug. 28, fol. 252	Inchoata de Cometis.	De cometis.
Hug. 32A, fol. 13	Mathematica varia mea. in quibus nihil fere alicujus momenti.	
Hug. 29, fol. 284	Rejecta ex Dioptricis nostris.	Rejecta ex Dioptricis.
Hug. 29, fol. 80	Adversaria ad Dioptricen spectantia. Inquisitio optimarum lentium si tantum figurae sphaericae culpa radij aberrarent quod secus est.	Adversaria circa Dioptrica.
Hug. 29, fol. 208	Dioptrica nostra descripta manu Niquetij.	
Hug. 32A, fol. 3	Adversaria eorum quae prior misi ad Mersennum, scilicet de motu acceleratio, de flexu Catenae. Crossed out: Item algebrae quaedam [effaced] imperfectae	Item eorum quae Hugenius puer ad Mersennum misit de motu accelerato, & flexu catenae.
	De Superficiebus Conoidum et Sphaeroidum.	De superficiebus conoidum, & sphaeroidum.
Hug. 26, fol. 19	De ijs quae liquido supernatant.	De liquido supernatantibus.
Hug. 31, fol. 254	Copie de mon Traité de la Lumiere, faite par M. Chasel.	Copia tractatus Hugenii de lumine.
Hug. 31, fol. 184	autre prieure [sic] Copie du Mesme.	

order and sometimes translated, as fasc. 3. Corresponding to a crossed-out title referring to letters involving l'Hospital, the description of fasc. 4 adds the information that the papers include Huygens' drafts of letters to Leibniz and others. Note that the side remark on Huygens' title reinterates the will's statement that these papers were stored separate from the others. Probably, he crossed the letters off his inventory of the third set and added the remark when he removed the letters from the set for further study. Already segregated from the main set by Huygens, it is not surprising that the group was assigned a separate codex when it went to the library. For fasc. 5 the catalogue expands Huygens' brief reference to 'various mathematics of others' into a very long list that continues for 63 entries, thereby implicitly justifying the split into still another codex. Fasc. 6 to 12 are the letters. Hence the only unmatched description in the catalogue is for fasc. 13. It is another

Table 2

Single Title in Cod. Hug.	Multiple Title List Hug. 32A, fol. 16	1716 Catalogue fasc. 2
Hug. 26A, fol. 90 [library cover]	De motu ex Percussione, descripta per Papinum.	De motu ex percussione. [in *Opuscula postuma*]
Hug. 26A, fol. 111 [library cover]	Item quae de occursu mollium corporum et semiduorum.	De occursu mollium, & semidurorum.
	Item adversaria posteriora ad hanc doctrinam spectantia.	Adversaria de motu per impulsum.
Hug. 26A, fol. 7	De motu ex Percussione [spilled ink obliterates a word (aut?)] [im]pulsu adversaria priora.	
Hug. 26A, fol. 6	De eodem adversaria [omnium] prima.	
Hug. 32A, fol. 6	Pieces touchant les Regles de la percussion.	Regulae percussionis.
Hug. 26, fol. 3	De vi Centrifuga.	De vi centrifuga. [in *Opuscula postuma*]
Hug. 28, fol. 176 [library cover]	De Automato Planetaris.	De automato planetario. [in *Opuscula postuma*]
Hug. 31, fol. 137 & fol. 161 [in pencil]	Derniere maniere d' expliquer les effets de l' Aimant pour Mr. du Hamel, et quelques brouillons du mesme sujet.	Ratio explicandi effectus magneticos.
	Microscopia.	Microscopia spectantia, inter quae epistolae N. Hartsoeker, & A. Leenwenhoeck.
Hug. 31, fol. 94	De Parelijs.	De pareliis. [in *Opuscula postuma*]
Hug. 29, fol. 268	Commencement de ma Dioptrique en Francois.	
Hug. 29, fol. 99	Item adversaria quaedam ad Dioptricen spectantia.	
Hug. 29, fol. 103	Dioptrice mea in qua nonnulla mutanda et addenda.	Dioptrica nonnulla. [in *Opuscula postuma*]
	Escrits de Mathematique dont j' ay donné Copie a l' Academie des Sciences a Paris.	De causis halonum, & pareliorum, De libella telescopica. De effectis perspicillorum. De barometro secundo & thermometro. Efficacia motus aquae, & aëris. Regula ad inveniendas curvarum tangentes. Demonstratio regulae de maximis, & minimis. De aequilibrio bilancis. Mechanica. Coronis circum Solem. Responsum ad objectionem Robervallii. Demonstratio de puncto reflexionis.
	Rescissa ex libris adversariorum	Rescissa ex libris adversariorum.

Table 3

Single Title in Cod. Hug.	Multiple Title List Hug. 32A, fol. 8	1716 Catalogue fasc. 3
	Commentationum Mathematicarum Recensio.	
Hug. 28, fol. 141 Hug. 28, fol. 158 Hug. 32, fol. 168	Mechanica artificia, et physica quaedam Glase. slypen. Niveau. Imprimerie. Pendule de poche. Instrumenten der Noorthollandes om canon de visschen.	Methodus vitra poliendi. [in *Opuscula postuma*] De horologio Hugeniano. [= pendule de poche] Nova chalcographiae ratio. Libella pro telescopiis. [= niveau] Machina ad ea quae in fundo maris sunt exantlanda.
Hug. 27, fol. 5	Musica adversaria.	Musicae adversaria.
Hug. 32A, fol. 11	Mathematica aliorum varia.	[see fasc. 5]
Hug. 32A, fol. 7	Epistolae Olai Romeri de mora Lucis cum responsis meis.	Epistolae O. Romeri de mora luminis, cum responsis Hugenii.
Hug. 32A, fol. 4	Literae Huddenij. cum nonnullis meis responsis.	Epistolae Huddenii cum Hugenii responsis.
Hug. 25, fol. 72	Paralogisme du Sr. du Laurent.	Paralogismus D. Laurentii.
Hug. 32A, fol. 5	Aengaende de Lengde vinding door mijn horologien.	De inventione longitudinis beneficio horologii Hugeniani.
		fasc. 4
	[Crossed out:] Correspondence avec le Marquis de l' Hospital [Not crossed out, to the side of the above:] dans un tiroir a part.	Epistolae Marchionis Hospitaletii, Leibnitzi, Mejeri, & de la Hire ad Hugenium, hujusque responsoriae, continentes varii generis Mathematicas demonstrationes, observationes Physicas, & reflexiones ad ea quae a Bernoullie actis Lipsiensibus sunt inserta.
		fasc. 5
		R. des Cartes epistolae ad J. Golium. Experimentum circa refractionem. De coronis circa candelas. *Les Mechaniques de Descartes.* V.F. Plempii epistola ad R. Descartes, cum objectionibus Fromondi, & R.Descartes ad illa responsis. Compendium Musicae R.Descartes. *Observations pour le mouvement, & repos de l'aymant.* Observata circa magnetem. [etc.]

long list that, disconcertingly, also appears to include work by others, such as 'Cassini's observation of Saturn', and so could be a continuation of Huygens' third set. Certainly, the agreement of the other descriptions with Huygens' three lists suggests that, although the 1716 catalogue was prepared after de Volder and Fullenius temporarily removed pieces from the collection for editing, it does accurately reflect how Huygens had his papers arranged at the time of his death.

The papers probably remained relatively untouched until a century later, when two separate events called attention to the collection. As part of a general appraisal of the library's holdings, P.J. Uylenbroek was asked to survey the Huygens papers. Uylenbroek's enthusiasm for what he found is endearing, and the modernity of his appraisal is astounding. He proposed publishing the manuscripts, and in the propectus for his project he stated that the papers would 'shed light on several obscure points of the history of mathematics'. Uylenbroek's description stressed the raw state of some manuscripts, calling them 'very rough drafts (vrai brouillons), but some precious rough drafts because they show us the first outlines of his ideas and the paths that led him to his remarkable discoveries'. In particular, Uylenbroek noted that the rough drafts of *De Motu* provided proof that Huygens planned to publish a large work on the subject, and he chided the earlier editors for not calling attention to these papers, especially since the manuscripts also bolstered Huygens' claim of priority with respect to Christopher Wren and John Wallis.[11] Briefly describing the letters, Uylenbroek voiced his regret that in general Huygens had not left Leiden the drafts of his replies, then expressed his delight that in the important correspondence over the calculus with Leibniz and l'Hospital the drafts were in the collection. Alas, his 1833 edition of these letters was the only fruit of his ambitious publishing plans.[12]

By a twist of fate, the bulk of Huygens' drafts entered the collection at almost the same time that Uylenbroek had noted their absence. A large portion of the family papers, which were still mostly intact after having been so devotedly amassed and preserved by Constantijn Lz, went on the auction block in 1813. Fortunately, much of the material at auction was purchased by the king; unfortunately, the purchase was then divided among three separate libraries.[13] Nonethe-

11 '. . . un riche dépôt de pièces authentiques, qui un jour pourront servir à répandre des lumières sur plusieurs points obscurs de l'histoire des mathématiques' [*Prospectus d'une édition des manscrits mathématiques de Chr. Huygens, légués a la bibliothéque de l'université de Leyde*, (4 pps, undated but the subscription list for the edition was open until May 1825), p. 1]. And '. . . ce sont de vrais brouillions, mais des brouillons précieux à cause qu'ils nous offrent les premiers projets de ses idées et les voies qui l'ont conduit à ses découvertes remarquables' [p. 3].

Dr. Eisso Atzema has researched Uylenbroek's relationship to the Huygens papers as part of his study on the role nationalism played in the creation of *Oeuvres complètes*. See his presentation 'A Genius True to His Country: Dutch nationalism and the forging of Christiaan Huygens as a scientific hero in the 19th century' at the conference 'Christiaan Huygens: 1695–1995', Leiden, July 7, 1995. He very kindly provided me with copies of the pertinent evidence that he has uncovered, in particular Uylenbroek's report to the curators of the library at Leiden (20 March 1823) and the prospectus.

12 *Christiani Hugenii aliorumque seculi XVII virorum celebrium exercitationes mathematicae et philosophicae. Ex manuscriptis in Bibliotheca Academiae Lugduno-Batavae* (2 vols. in one, The Hague, 1833).

13 For a history of the family papers, see A.M.Th. Leerintveld, ' "Ter goeder memorie van mynen naem;"

less, Leiden did receive all papers identifiably Christiaan's, including the packet of drafts, for which he had long before written the cover 'Epistolae meae' (Hug. 32A, fol. 2). The first addition to the Codices Hugeniorum, however, actually came from A. J. Royer, one family member who, instead of selling, donated his share directly to the library.

Although the library had to assimilate substantial additions to the collection, the original papers do not appear to have been disturbed from their previous arrangement.[14] An 1852 catalogue of letters at Leiden, which admittedly has some major misattributions regarding provenance, merely joins drafts and other papers written by Christiaan to the end of its list of the letters to Christiaan already in the collection and, then, lists separately letters to his father that the Royal donation also contributed.[15]

Some time later, most likely after 1869, a composite catalogue of the entire Codices Hugeniorum was pasted together for internal use from cut out portions of the 1716 and 1852 catalogues. In other words, the descriptions of the original codices were transferred without alteration and I presume that the papers were likewise unaltered. The codices did receive new shelf numbers, because the division into books and bundles of loose sheets was eliminated and replaced with a single sequence of numbers, which at that point ran to 39. In particular, the bundles of loose papers were assigned numbers 25–29 and 31, with all the letters of bundles 6 to 12 collected into codex 30. The papers *had* suffered some mishandling in the first 150 years of their Leiden residency. Fullenius and de Volder wrote on them, and some curious soul, in order to expose the original passagés underneath, shattered the sealing wax that glued large corrections over their respective sites. But, I would claim, the papers originally bequeathed to the library were still in the order in which Huygens had arranged them.

This state of benign neglect changed dramatically when work on the *Oeuvres complètes* began in the 1880s.[16] The manuscripts were taken from Leiden, repackaged by the senior editor, and parcelled out to various scholars working on the project. I have found no evidence to show that the editors catalogued the loose

de nalatenschap van Constantijn Huygens', in V. Freijser (ed.), *Soeticheydt des Buyten-levens: Leven en Leren op Hofwijck* (Delft, 1988), pp. 97–115.

14 Some time after the Royal donation of 1823, someone added brief listings for fasc. 14–21 in UB Leiden's copy of the catalogue; the listings correspond to the Royer (fasc. 14) and Royal donations. In particular: fasc. 14 is listed as papers concerning Christiaan Huygens; fasc. 15, letters between family members; fasc. 16, Christiaan to and from Constantijn Jr; fasc. 17, the manuscript of Cosmotheros and letters of Chieze; fasc. 18, minutes of letters of Christiaan to known and unknown scholars; fasc. 19, two bound folios of copies of letters [made by Constantijn Lz]; and fasc. 20–21 letters to Constantijn Huygens Sr.

15 Jacob Geel (ed.), *Catalogus Librorum Manuscriptorum, qui inde ab anno 1741 Bibliothecae Lugduno Batavae accesserunt* (Leiden, 1852), pp. 280–86. The Geel catalogue has errors: the letters of the original bequest (fasc. 6–12) are included and attributed to the Royer donation of 1809, while the Royer donation (fasc. 14) is not described; the letters from the Royal gift (fasc. 15, 16, 20, 21) are listed as having been donated in 1822.

16 Edited and published under the auspices of the Holland Society of Sciences (Hollandsche Maatschappij der Wetenschappen), *Oeuvres complètes de Christiaan Huygens* (The Hague) began appearing in 1888 and totaled 22 volumes when finished in 1950.

folios,[17] recorded to whom each packet was assigned, or codified in writing the procedures by which the papers were to be handled, let alone edited. The unintended consequence of this neglect was the obliteration of Huygens' arrangement.

An example reveals the disintegration. The original bundle of loose sheets catalogued as fasc. 2, had by then been renumbered to 26. Note that Huygens had listed six titles that pertained to his studies on the motion of colliding bodies [Table 2]: the treatise *De Motu*, papers on collision involving elastic bodies, later papers, earlier papers, the very first papers, and 'les Regles de la percussion'. The 1716 cataloguer reduced the list to four, gathering all the 'adversaria' into one and translating the last title into Latin. Editor D.J. Korteweg divided the papers on motion into *eight* packets, which he lettered *A* to *H*.[18] *A* might correspond to Huygens' 'first papers' and *B* to Huygens' 'earlier papers'. *C* to *E* contain manuscripts used by Papin to copy out the treatise. The apparent reason for the tripart division by Korteweg is that *E*, a lone double folio, is a copy by another amanuensis of part of *C*, which relates to the version sent to London in 1669. Korteweg's *F* holds the cleanly copied treatise, while *G* includes papers on elastic collision. All of these packets, arranged in Korteweg's order, are no longer in codex Hug. 26 but have been segregated into Hug. 26A, a new codex created especially for them.

How extensive was this reordering? There are now more codices and divisions, but perhaps the arrangement of the papers within the packets has remained unchanged. After all, Korteweg made eight divisions, but two were rather spurious, thus it could be argued that he really only returned to Huygens' six. Indeed, at first glance, there seems to be a one to one correspondence between their descriptions, except that Korteweg reversed the sequence of the packets in order to make them chronological. But are Korteweg's divisions *C* to *E* the same as Huygens' group with the French title, so named perhaps because they were copied in Paris? If so, where are Huygens' 'later papers'? And what about packet *H*?

'Where is that bundle?' Korteweg asked J.A. Vollgraff on 12 Jan 1928. He knew that E.J. Dijksterhuis had some of the manuscripts, but not the *H*s, which seem to have gone missing.[19] Not to fear, they were found. And none too soon, for Korteweg had retired and the manuscripts were being returned to the library.[20] An internal library note, made just one day after that urgent demand, recorded the deteriorated state of the manuscripts, including the fact that ink had been spilled down the center of workbook *A*, insisted on the immediate return of all remaining manuscripts, and decreed that all work must henceforth take place in the library.[21] H.A. Lorentz, who was the current president of the commission overseeing the

17 The library catalogue seems to have been considered sufficiently detailed. Indeed, perhaps this is when the composite list was actually made by the library.

18 His list is now Hug. 26A, fols. 1–2.

19 'Waar is dit bundelije' [Museum Boerhaave, Vollgraff Archives, Arch. 192, second box, first bundle].

20 See Vollgraff's appreciation of Korteweg in *Oeuvres complètes* vol. 21, 892–893.

21 See the note dated 13 January 1928 in 'Verslag van de Bibliothecaris aan Curatoren over de jaren 1925/26 t/m 1946/7'. Another note, dated 10 October 1928, refers to the belated return of a packet of drawings obviously belonging to the collection.

edition and who apparently was the possessor of our wandering *H*, as well as part of *G*, asked that the whole bundle be left as a separate entity. When he died only a month later, Vollgraff, the new senior editor, fulfilled his request, thus creating codex Hug. 7A.[22] What made these papers so very interesting? Korteweg had pulled from the collection all those papers that dealt with the question of absolute and relative motion. Quite clearly, then, the papers had been disturbed at the local level. Thus, the first folio in packet *H* begins with notes regarding Borelli's work on circular motion. But surely those should be in Huygens' next section on centrifugal motion? And where did *De Vi Centrifuga* go? Surprisingly, it and its title page stayed in Hug. 26, although the *Oeuvres complètes* refers to it as if it were a free-standing entity. But it is followed in Hug. 26 by the treatise on floating bodies that Huygens and the 1716 catalogue assigned to the papers that should be in Hug. 25. Because the papers contained in the rearranged packets do pertain to questions of motion, it is impossible to ascertain how much of Huygens' arrangement remains. In particular, although most of his papers on relative motion postdate his reading of the *Principia* and thus are definitely later than the material copied in the 70s, we cannot presume that Korteweg's *H* is a faithful preservation of Huygens' bundle 'later papers'. Certainly, Huygens did not leave us with a title page for a bundle on relative motion.

Moreover, what are we to make of a title page that refers to 'mathematical pieces of mine, few of any importance?'[23] A variant is repeated in the list of titles that begins with Alhazen's problem [Table 1, item 3], although it is not repeated in the 1716 catalogue. Alas, I cannot tell you what mathematical papers Huygens thought were insignificant, because the cover no longer resides with the papers to which it was referring. It, the three lists of multiple titles that have been the focus of my discussion, and ten other pages with single titles, are sequentially stored (fols. 2–16) in a codex (Hug. 32A) that was not created until after the great additions of the nineteenth century. In fact, the bulk of the codex is material purchased at auction in Paris in 1884.

Vollgraff, in his report on the creation of codex Hug. 7A, confirms our worst suspicions: 'D.J. Korteweg of Amsterdam, who for a long time had the manuscripts of Huygens at his disposal, generally rearranged the loose folios. . . . Moreover all the loose folios are henceforth preserved as they were arranged in Amsterdam.'[24] Potentially, everything could have been moved! Most definitely,

22 An account of this rearrangement is given in *Oeuvres complètes* vol. 21, 415, n. 1. Instead of assigning the codex the number 26B, thus keeping the papers somewhat sequential to their original neighbors, Vollgraff designated them *L* in imitation of Huygens' lettering of his workbooks. (Besides the chronological *A* to *I*, *K* was a compendium of material taken from the previous nine workbooks.) Thus, the packet of loose papers was placed forward in the collection with the bound books. The codex has been edited, including passages not in *Oeuvres complètes*, by Gianfranco Mormino in his dissertation for the University of Milan, *Penetralia Motus: La fondazione relativistica della meccanica in Christiaan Huygens, con l'edizione del 'Codex Hugeniorum 7A'* (Florence, 1993).

23 'Mathematica varia mea pauca alicujus momenti' [Hug. 32A, fol. 13; edited to *Oeuvres complètes* vol. 20, 195].

24 'D.J. Korteweg d'Amsterdam . . . qui a longtemps eu les manuscrits de Huygens à sa disposition, a généralement réarrangé les feuilles détachées. . . . D'ailleurs toutes les feuilles détachées sont désormais conservées comme elles avaient été arrangées à Amsterdam' [*Oeuvres complètes* vol. 21, 415, n. 1].

the title pages were moved. Which, among other consequences, means that Huygens' lists, in which he meticulously recorded his arrangement of his papers, enumerate mere titles, detached from any underlying content that their accompanying papers would have provided and, thus, devalued. The 1716 catalogue, now discovered to be an accurate reflection of Huygens' own organization, is likewise rendered contentless and thus relatively worthless. In a further disintegration of the collection, all of the editors' notes, primarily Korteweg's, are now interleaved with the manuscripts. Thus, when the folios of these codices were numbered by the library, probably soon after being returned in the panic of 1928, the editors' notes were incorporated into the foliation.

The letters, also being loose, were thus also moveable. Originally some letters were housed with related manuscripts: Uylenbroek edited the Leibniz and l'Hospital correspondence that he found among the calculus papers; the 1716 catalogue repeated Huygens' inclusion of correspondence with Olai Romer and Johan Hudde in with his technical inventions; and the catalogue even expanded his entry 'Microscopia' with the information that letters from Nicholaas Hartsoeker and Antoni van Leenwenhoek were in that section. Still, most of the letters to Huygens that went to the library in the original bequest were arranged alphabetically by sender in the six bundles assigned 6–12 in the 1716 catalogue. The Royal donation brought into the collection the bulk of the drafts and some family letters, as well as a very large set of letters to Constantijn Sr (now Hug. 37). Today, all letters involving Christiaan are stored separately in three alphabetical runs under the same codex number, Hug. 45: letters written by persons other than members of the family, primarily consisting of letters to Christiaan Huygens; letters between members of the Huygens family, but not including Philips Doublet, his cousin and brother-in-law; letters written by Christiaan to others, primarily consisting of his drafts. Note this very careful description! Third party letters and copies are in the first section. Originals sent by Huygens, including those obtained by the library in later acquisitions, are interleaved with the drafts in the third section. For example, there are many doubles of letters to Frans van Schooten housed together. This configuration causes identification problems, because some letters in the third section show physical characteristics suitable to originals, such as fold lines and addresses, but might be only drafts or even unsent letters.

But manuscripts do not have to move in order to be displaced. Proof by example: *De Vi Centrifuga*. The autograph manuscript remains in codex Hug. 26, but Fullenius and de Volder rearranged the text in their edition in accordance with a list of thirteen propositions presented without proofs in part 5 of *Horologium Oscillatorium*. Obviously wishing to lighten their load by building on work done before them, the editors of *Oeuvres complètes* reprinted the previous edition (as they did for five of the six works in *Opuscula Postuma*) and, thus, repeated the reordering of the text. The complete autograph is a set of loose sheets, paginated by Huygens 1 to 28, which he cut from two different places in workbook *A*. The beginning pages of the manuscript were not edited by Fullenius and de Volder, so we have only the modern edition's version, which is labelled a draft and is appended to the reprint of the main text. A comparison of the manuscript with the hybrid product of two sets of editors confirms the addage about too many cooks. As was their custom, the editors of *Oeuvres complètes* arbitrarily divided the draft into sections and suppressed a long calculation as well as preliminary experimental values. Moreover, the printed figures are not facsimiles of Huygens' originals, although they might be mistook for such because they are hand-drawn imitations.[25] The transcription of the draft ends prematurely, because Fullenius and de Volder appropriated the last three propositions plus some commentary for their edition. In fact, the commentary that begins *De Vi Centrifuga*, as edited by Fullenius and de Volder, is the true end of the draft. At the bottom of that page, Huygens goes on to study fall along inclined planes, and this analysis continues onto pages that are still in the workbook. His cut-out pages resume with a failed attempt to compare fall along arcs using their tangent planes.[26] In order to duplicate the list of propositions in *Horologium Oscillatorium*, Fullenius and de Volder not only lifted propositions from the first half of the manuscript but also provided their own proofs when a proposition inconveniently did not appear in the manuscript at all.

Is the edited *De Vi Centrifuga* a treatise by Huygens? He did not reorder the manuscript, insert additional proofs, and have it copied – as, we shall see, he did for *De Motu*. Nor did he acknowledge its existence in his will. In fact, Fullenius and de Volder felt the need to justify what they had done, particularly defending their added proofs as in the style of Huygens.[27] The editors of *Oeuvres complètes* point out that the proofs are not that Huygenian. Nonetheless, when it came time for them to deal with this manuscript, they reproduced the rearranged edition. Moreover, they added a facing-page French translation, which they did with all major treatises. The preliminary work, now relegated to an appendix, only appears in Latin.

Ironically, the very features of their treatment that have disturbed me also unsettled Korteweg. He worried about the drawings and asked why only the main portion was given a French translation. Most of all, he fretted about whether they had clearly distinguished what was by Huygens and what was by de Volder and Fullenius.[28] Yet, after all, Huygens did record the list of propositions that form the outline for the edition by de Volder and Fullenius (Hug. 13, fols. 22–23). He even gave the loose set a title page: 'De Vi Centrifuga' (Hug. 26, fol. 3).

Given their penchant for marking up the texts that they handled, I feel confident claiming that the pagination on the autograph of *Descriptio Automati Planetarii* is by Fullenius and de Volder, not by Huygens as the modern editors presume. This was the second work not specified in the will that Fullenius and de Volder edited. Their edited text continues on beyond the last page of the autograph. Where did that material come from? The text probably would not fill more than one extra

[25] Of course, during the early years of the project they were working before photographic plates were available. In fact, I do not know how they produced most of their early figures. In the case of some letters in facsimile, the originals are now missing from the collection.

[26] For a complete reconstruction, see my *Unrolling Time: Christiaan Huygens and the mathematization of nature* (Cambridge, 1988).

[27] Preface to *Opuscula postuma* and *Oeuvres complètes* vol. 16, 238–9.

[28] Museum Boerhaave, Vollgraff Archives, Arch. 192, second box, first bundle, 5 Dec 1927.

folio. Curiously, the autograph ends with most of the last page blank. Perhaps the last part was an addendum on a separate sheet that subsequently was lost. I have not found it elsewhere in the manuscript collection. But perhaps it was created by Fullenius and de Volder from values given elsewhere in Huygens' papers, just as they did with the extra proofs for *De Vi Centrifuga*.[29]

Even single manuscript pages have not been immune from editorial rearrangement. I will confine myself to describing the fate of just one, fol. 11 of Hug. 31. Although there are lines between the paragraphs, the text looks as if it were written at one sitting. However, the editors of the *Oeuvres complètes* broke it apart according to topics, which is the way the whole edition is arranged. The second and third paragraphs, because they are about music (the tone of a flute, the ringing of a bell), went to vol. 19, p. 377; the fourth, on colliding bodies, to vol. 16, p. 168; the first and fifth, on the effect of heat on substances, were not edited or even noted. Why did Huygens record these passages on the same page? An intriguing question about a textual relationship is likely to remain unasked when the underlying physical relationship is thoroughly obscured by the way it is edited.

Moreover, as much as they can be, Huygens' loose manuscripts are now also physically reorganized by scientific fields: codex Hug. 25 comprises mathematics, 26 mechanics, 27 music, 28 astronomy, 29 dioptrics and 31 physics. The editors of *Oeuvres complètes* even refer to them that way in the later volumes: Chartae Mathematicae, Chartae Mechanicae, etc. Surprisingly, the volumes of *Oeuvres complètes*, which are arranged by the editors according to disciplines, do not sequentially reproduce the papers, even though the papers were also arranged by the editors according to disciplines. At least the editors' classification scheme is self-evident.

I have to admit that Huygens' original arrangement is quite puzzling. It does not appear to be based on any taxonomy or theory of knowledge. I conjecture that Huygens sorted his papers specifically in preparation for death, and that he did so twice. We know from a letter by Francis Vernon to Henry Oldenburg that in 1670, when Huygens was ill and afraid that he was dying, he entrusted sections of his papers to various people for safekeeping. In particular, he gave Vernon a sealed packet containing the proofs of twelve propositions on motion that he had previously sent to the Royal Society as anagrams and a larger, unsealed packet containing 'some propositions about motion rough cast & in their first draught & on the Paper which wraps them up is written on the outside de Motu per impulsum'. If Huygens died, Vernon was to give the first set to the secretary of the Académie des Sciences and the second set to the Royal Society, most particularly to Lord Brouncker, 'whom hee said hee knew to have a very cleare insight into the Doctrine

29 The manuscript is in Hug. 28, fols. 177r–188v. The treatise is reproduced in *Oeuvres complètes* vol. 21, 589–639. The later editors casually mention a missing end on p. 581 and even detail in their n. 3 places where the earlier editors made changes in the text, but they do not consider the possibility that the end is likewise fabricated. Note that the extra theorems made by Fullenius and de Volder for *De Vi Centrifuga* are not inserted into the manuscript collection. Consequently, any additions by them to *Descriptio Automati Planetarii* would probably not be there either.

of Motion & whom (hee was well assured) did understand those Points by what hee had seene by him already & hee hoped that for his sake & upon his request hee would digest those Papers which were confusedly written & which hee had not time to finish soe that if hee judged them worthy of the world the world might bee the better for them'.[30] Vernon's report sounds almost like a draft of Huygens' will, with Brouncker nominated for the role, at least with respect to *De Motu*, that de Volder and Fullenius would eventually take.

It seems plausible to expect that, once he had recovered and had returned to Paris, Huygens himself might finish some of those 'confusedly written' rough drafts rather than risk a fatal replay that would leave his worldly worth in the hands of another. For example, surely this encounter with death hastened the long-awaited publication of *Horologium Oscillatorium* in 1673, although court politics undoubtedly influenced the event as well. Moreover, sometime between 1673 and 1675, Huygens arranged in sequence his papers on motion for Papin to copy.[31] That is, the 'autograph' of *De Motu* was not a complete draft, roughly written but proceding from beginning to end; rather it consisted of parts taken from at least three previous drafts plus material added at the moment. To guide Papin through this maze, some pages have instructions, including the delightful 'this ought to follow what the following sheet holds'.[32] It is remarkable that Huygens had never made a clean copy of the whole before, especially during 1668–69, when he was disputing his priority over the laws of motion with the English. It is more remarkable that, although the clean copy contained the text precisely as it would eventually appear, for the next twenty years Huygens never published it.

Although Huygens was not lucid in his last days, he did have enough physical warnings near the true end to have organized his papers once more for posterity. Certainly the final arrangement as listed by Huygens and catalogued by the library postdates 1687, because one of the entries refers to papers sent in that year to Philip de la Hire for publication by the Académie des Sciences.[33] In fact, because of the inclusion of *Traité de la lumière* on a different list, I presume that the lists were created later than 1690, when it was published. Probably during this reorganization of his papers, Huygens cut from his workbooks studies that he thought were particularly important. *De Vi Centrifuga* is one, but note that the second list

[30] Dated 25 Feb 1670; Royal Society, Early Letters V.12. Published in *Oeuvres complètes* vol. 7, 7–13, and Oldenburg, *Correspondence*, vi, 501–7. There is no title page 'de motu per impulsum' in the Codices Hugeniorum. However, there is a 'de motu per impulsum adversaria priora' (Hug. 26A, fol. 7), which would pertain to the papers given Vernon once Huygens had completed *De Motu*.

[31] The copy falls after 1673, because it cites the *Horologium Oscillatorium*, and before August 1675, by which time Papin is writing from London (see *Oeuvres complètes* vol. 7, 490–1). Another factor that might have prompted the copy is the publication in 1673 of Edme Mariotte's *Traitté de la Percussion ou Chocq des Corps*, which Huygens felt should have cited him.

[32] 'Hic sequi debent quam folio sequente continentur' [Hug. 26A, fol. 45v]. It is possible to reconstruct Huygens' entire sequence by following his signs and notes.

[33] See the letters between de la Hire and Huygens in *Oeuvres complètes*, vol. 9, 91–2 and 95–7. Not all of the items sent by Huygens appeared in *Divers ouvrages de Mathématique et de Physique. Par Messieurs de l'Académie Royale des Sciences* (Paris, 1693).

includes a section specifically for such pieces ('Rescissa'). Yet he still bequeathed the workbooks themselves, and his lists include many other loose 'adversaria'. Hence, I do not think that he rearranged his papers so that he could withhold items. Rather, he seems to have been directing attention along certain primary paths through a vast unknown. As the 1670 incident involving Vernon suggests, he appears to have focused on items that would be 'worthy of the world', that is, publishable.

The manuscript collection contains no prepublication drafts, either in Huygens' hand or as copies, of the ten largest works printed in his lifetime. Apparently, the final version set by the printer was thrown out. Thus, the source material for such major works as *Horologium Oscillatorium*, *Systema Saturnium*, and *Discours de la cause de la Pesanteur* is limited to the early development of the treatise, not its ultimate reworking.[34] The only exception to this rule is *Traité de la lumière*, probably because Huygens had originally intended it to be the first chapter of his *Dioptrica* and retained it in order to reattach a Latin version of it to the complete work. As I've noted, *Cosmotheoros* was in press at his death. Evidently, his brother kept the manuscript after it was printed, for it only entered the collection as part of the Royal donation.

A careful look at Huygens' lists of the remaining manuscripts reveals that most of the titles on one sheet concern items that he was entrusting to others to publish, including the papers sent to de la Hire in 1687 [Table 2]. Thus, it is not such a coincidence that all but one of the items edited by de Volder and Fullenius are listed together, first by Huygens and hence by the 1716 catalogue. Perhaps, it is more to be wondered why the treatise on grinding lenses specified by Huygens in his will is not grouped with the others. Moreover, perhaps de Volder and Fullenius should not be too severely censored for extending their editorial mandate to two items that are, afterall, also included in this set. Another list appears to be items that either he had rejected, including older drafts and juvenilia, or he had already published, including work on Alhazen's problem and the *Traité de la lumière* [Table 1]. Fortunately, they were not thrown away as the final drafts of the earlier published works had been. Mixing technological projects with mathematical discourses, the third list challenges my attempted division according to publication status [Table 3]. Most are not treatises, not even in a rough draft form. One unifying thread is that most pieces appear to have involved other people, from actual papers of someone else to intellectual give-and-take communicated in letters to items subject to priority disputes. Taken as a whole, they situate Huygens within the scientific community and, thus, establish his worth. I suspect that he wanted his editors to cite the material in an historical overview of his achievements that would emphasize his priorities, as indeed Uylenbroek was to do in his assessment. Certainly the three lists do not classify his papers by an intrinsic relationship based on subject matter. Is the division Publish, Ignore, and Cite? Although convenient

34 Also in the list are *Theoremata de quadratura*, *De circuli magnitudine*, *De Saturni luna observatio nova*, *Horologium* of 1658, *Kort onderwijs*, *Van Rekeningh in Spelen van Geluk*, and *Astroscopia compendiaria*.

for editors making a posthumous collection, this is a strange way to keep your working papers.

The obvious question: had Huygens always kept his papers ordered by publication status, at least after the 1670s, or did he rearrange them specifically for his editors? Indications favor the latter choice. For example, there are places in his manuscripts where he refers back to research recorded in an earlier workbook. Thus, he was not habitually tearing out pages and storing them loose. In addition, the condition of the autograph for *De Motu* is very puzzling. Those pages are no longer arranged in the sequence that Papin was to follow. Instead, they have been returned to their positions in the earlier drafts. Who moved them? Korteweg, with his three divisions *C*, *D*, *E*? But the editor who wrote the introduction to *De Motu* in *Oeuvres complètes* seems to have been unaware that they possessed the autograph. So, most likely Huygens himself disturbed the sequence, as evidenced by his six divisions. But why did he rearrange them and when – immediately after they were copied or as a final preparation for posterity? I have conjectured elsewhere that he pulled the papers apart in order to use them in a broader study that involved *De Vi Centrifuga*, which would also explain why he did not publish *De Motu* in the 1670s.[35] My conjecture might be true, but the current arrangement does not provide overwhelming evidence. However, who is most responsible for the current arrangement? I suspect that my evidence would have once rested behind the cover with the title 'later papers pertaining to this treatise', which is not one of Korteweg's divisions. What is by Huygens and what is by his editors? Surely I have proved my claim that the Codices Hugeniorum preserve the work of both Christiaan Huygens and his editors.

35 See my article cited in n. 1.

7

The Archive and *Consulti* of Marcello Malpighi: some Preliminary Reflections

Domenico Bertoloni Meli

Introduction: a brief survey of Malpighi's manuscripts

The survival and present-day location of Malpighi's manuscripts reflect the pattern of dissemination and publication of his writings during his lifetime and immediately following his death. The Royal Society, for example, owns manuscripts of several works from *De Bombyce* to *Anatome Plantarum*, as well as many letters by Malpighi. The largest portions of Malpighi's manuscripts, however, are preserved in several libraries in Italy. Among the most conspicuous holdings are those of the Biblioteca Lancisiana at Rome, founded by Malpighi's friend and colleague Giovanni Maria Lancisi, and the Library of the Medical Faculty at Florence University, containing Malpighi's letters to the Florence anatomist Lorenzo Bellini.[1] The most extensive collection is housed at the University Library and other libraries in Bologna, notably the Biblioteca Comunale dell'Archiginnasio. These amount to several thousand folios of manuscripts.

The great bulk of the manuscripts at the Bologna University Library was discovered in 1830 by a school teacher, Gaetano Atti, then working at Crevalcore, Malpighi's birthplace. On Malpighi's death his manuscripts had been entrusted by his brother, Bartolomeo, to his favourite disciple and nephew Ippolito Francesco

In writing this essay I benefited from several conversations with Marta Cavazza, Silvia De Renzi, Roger French, and John Henderson. I thank them all for their help. I am also grateful to Michael Hunter for his helpful comments. I have adopted the following abbreviations:

Consulti	*Consulti di Marcello Malpighi (1675–1694)*, ed. G. Plessi and R.A. Bernabeo (3 vols., Bologna, 1988–92)
Correspondence	*The Correspondence of Marcello Malpighi*, ed. H.B. Adelmann (5 vols., Ithaca, 1975)

1 Catalogues and brief descriptions of Malpighi's manuscripts can be found in G. Atti, *Notizie edite ed inedite della vita e delle opere di Marcello Malpighi e di Lorenzo Bellini* (Bologna, 1847), pp. 457–86; C. Frati, *Bibliografia delle opere a stampa di Marcello Malpighi e degli scritti che lo riguardano*, originally published in [U. Pizzoli (ed.)], *Marcello Malpighi e l'opera sua* (Milano, 1897, reprinted London, n.d.), pp. 4–19; H.B. Adelmann, *Marcello Malpighi and the Evolution of Embryology* (5 vols. [continuous pagination], Ithaca, 1966), pp. xvii–xix; C. Dollo, 'Gli inediti malpighiani', in G. Canziani and G. Paganini (eds.), *Le edizioni dei testi filosofici e scientifici del '500 e del '600* (Milan, 1986), pp. 209–20. See also L. Miani-Belletti, 'Autografi di Marcello Malpighi alla Biblioteca Universitaria di Bologna', *Accademie e Biblioteche d'Italia*, 49, n. 6, 1981, pp. 438–40; Ministero per i Beni Culturali e Ambientali, *Antichi e Rari* (Turin, 1991), pp. 113–15; D. Bertoloni Meli, 'Additions to the Correspondence of Marcello Malpighi', in D. Bertoloni Meli (ed.), *Marcello Malpighi, Anatomist and Physician* (Florence, 1997).

Albertini, also from Crevalcore. When he died in 1738, they passed to his heirs, the Pigozzi family. It was in their house that Atti was shown, to his astonishment, letters by Henry Oldenburg, Giovanni Alfonso Borelli, and Lorenzo Bellini, drawings of plants and anatomical parts, and drafts of Malpighi's works. Atti provides us with a vivid picture of his excitement and, more interestingly, clues as to what happened to the manuscripts:[2]

> Then warm with happiness I made the relative of the owner collect all the papers, which were used in that house for the most humble tasks, and arranged that they be classified according to subject. This task was given to me, as the Discoverer of the manuscripts. I accomplished it with the greatest pleasure and arranged them in seventeen substantial volumes, put together to the best of my abilities. Their order and contents are clarified in the following Catalogue.

This quotation suggests that an unspecified portion of Malpighi's manuscripts was lost because of the carelessness of their owners. Early in 1684 Malpighi's house suffered a fire and many of his papers went up in flames, thus that date marks a line before which relatively fewer texts have survived. By contrast, the damage inflicted on them by the Pigozzi family appears to have been random. Further, it is clear from the quotation above that the present arrangement of Malpighi's manuscripts reflects Atti's choices. Unfortunately, it is not always easy to determine the extent of his interventions and consequently to gain a picture of Malpighi's original arrangement or Albertini's, although some suggestions can be put forward.

In some cases we are lucky enough to be left with what appears to be relatively unproblematic material, such as a small manuscript volume with the autograph title *Anatomica*, where Malpighi annotated in alphabetical order most of the autopsies he performed or witnessed between 1666 and 1693. Together with post-mortem reports, Malpighi included annotations and summaries on anatomical matters, such as excerpts from books. For ease of reference, he glued on the right margins of successive pages small labels with the letters from A to T. The order given by Malpighi to this material is of great interest in showing how he classified the various cases, i.e., generally according to the name of the disease, starting from *apoplexia*, though at times he referred to the name of the organ affected, such as *pulmones*. Unfortunately, from the first edition of that text by Ladislao Münster it is not clear under which letters the various reports were filed and this cannot be easily inferred. Sadly, the more common and more accessible translation by Luigi Belloni complicates the matter further, because he rearranged the whole material chronologically and omitted all the annotations falling outside the category of

2 Atti, *Notizie* (n. 1), p. 456: 'Caldo allora di allegrezza feci raccogliere al parente del proprietario le carte tutte, le quali servivano in quella casa pei ministeri pi vili, e procacciai, che ne fosse fatta una classificazione per ordine di materie, il qual carico venne a me, come Ritrovatore, affidato. Ci eseguii con un compiacimento che mai il maggiore, e formai con questi scritti 17 volumi ben pingui, messi in sesto come si pot meglio, dellordine, e delle materie de quali se ne chiarisce chi scorre il Catalogo, che riporto qui sotto.' See also Aldelmann, *Embryology* (n. 1), pp. 665–8.

post-mortem reports. Yet it is of interest that Malpighi filed together his own observations and reports from the literature.[3]

A companion volume containing an alphabetical list of excerpts from the medical literature is structured in the same way, namely with small labels with the letters A to T glued to the right margins of successive pages. The list is not based on the names of the authors excerpted, but rather on the illnesses and parts of the body. It includes entries such as A, *arthritide*, B, *bile*, D, *dolor, diaeta*, E, *epilepsia*, F, *fluxio, febris, flatus*, I, *ittericia*, down to P, *pericardium*, R, *respiratio*, S, *sputatio, spiritus*, etc. The organisation of this private material should be of interest both to the historian of seventeenth-century archives and to the medical historian attempting to reconstruct how information about organs and diseases was classified by Malpighi.[4]

Among the most important manuscripts still awaiting a critical edition is the so-called diary, volume II of MS 2085, an extensive though unfortunately incomplete (thanks to the *famiglia* Pigozzi?) set of dated miscellaneous observations and reports on illnesses and therapies, anatomical observations, public anatomies held at the Bologna theatre, and conversations held with visitors such as Nicholas Steno and Gottfried Wilhelm Leibniz. Thus this document presents a fascinating cross-section of Malpighi's activities as physician, anatomist, and university professor. This manuscript includes also a few post-mortem reports, partly published by Münster.[5] Most of them were later copied in *Anatomica* (MS 2085, XII), thus indicating Malpighi's special interest in them. Although under the same classmark one finds several loose sheets, the largest part of Malpighi's diary consists of a batch of folded folio sheets inserted one inside the other. It is clear that Malpighi did not write the folio sheets separately and then put them together, but rather he first arranged the papers in this way and then wrote on them as on the pages of a book, filling first all the left sides and then moving to the right ones. Thus the physical appearance of the manuscripts shows that from early on Malpighi had planned to keep extensive records of his researches and intellectual pursuits for later use. The index of a small hand drawn by Malpighi on the left margin, serving the purpose of our *Nota Bene*, points to some passages of particular interest. In 1978 Luigi Belloni announced that Howard Adelmann and himself were working at an edition of volume II, though apparently this important work could not be brought to completion.[6]

Malpighi's university lectures too, preserved in the largely unexplored volume

3 Biblioteca Universitaria, Bologna, MS 2085, vol. XII. L. Münster, 'Marcelli Malpighii anatomica sive in cadaveribus sectis observationes', *L'Archiginnasio*, 60 (1965), 170–228. L. Belloni (ed.), *Opere scelte di Marcello Malpighi* (Torino, 1967), pp. 411–48. I shall return below to the problem of the editors' criteria for publishing a text.

4 Biblioteca Universitaria, Bologna, MS 2085, vol. XIII: 'Repertorium autographum per alphabeti ordinem quo plures ex variis auctoribus excerptae sententiae de rebus medicis annotantur.' To my knowledge this manuscript has not been studied.

5 L. Münster published without comment in 'Anatomica' some of the reports in MS 2085, II, which were not transcribed by Malpighi in MS 2085, XII.

6 L. Belloni, 'Marcello Malpighi', in G. Arrighi et al., *La scuola galileiana. Prospettive di ricerca* (Florence, 1979), pp. 137–53, on pp. 152–3; Dollo (n. 1), p. 211.

III of MS 2085, contain material of considerable interest on his roles as university professor and neoteric anatomist. Lastly, I wish to mention Malpighi's *consultationes* or *consulti*, several volumes of which are preserved at Bologna, Rome, and elsewhere both in his own hand and in copies made by his disciples and other physicians. I shall discuss this material below.

Before his death Malpighi arranged for many of his manuscripts, notably his *Vita*, or autobiography, and replies to Michele Lipari and Gerolamo Sbaraglia, to be translated into Latin by his friends Silvestro Bonfiglioli and Giovanni Lodovico Donelli and published posthumously by the Royal Society. Since that edition is notoriously inaccurate, readers may consult the draft preserved at the Bologna University Library both in order to study cancellations, and to correct the many misprints. The draft, for example, shows the transition from a brief preliminary version of the autobiography in Italian, first published at the beginning of our century, and the final version dispatched to London. Moreover, a puzzling reference to 'Cartesius' in the London edition, which had deceived several readers and was first tentatively corrected by Corrado Dollo, can now be definitely read as 'Cortesius', i.e., the Bolognese anatomist Giovanni Battista Cortesi.[7]

By contrast to those manuscripts, Malpighi forbade his disciples to publish his *consulti*, arguing that his extensive anatomical investigations had prevented him from devoting himself sufficiently to practice, a remarkable claim by the pope's chief physician.[8] Whereas his anatomical writings and papers were in his possession and could be entrusted to followers and disciples, the *consulti*, by their own nature, were sent to a wide range of patients and physicians over whom little or no control could be exerted. It appears that at least one of the reasons why Malpighi was concerned with the ownership and fate of his manuscripts was to secure for posterity that his image would remain associated with his anatomical works rather than medical practice. Indeed, despite the publication of some of his *consulti* in 1713, 1743–5, 1747, and in our century,[9] not only the fame but also our knowledge of Malpighi remain largely based on his anatomical findings. Hence the mutual reinforcement between preference for theory and publication criteria. A less radical dichotomy between theory and practice and a more careful analysis of the entire corpus of his surviving manuscripts, including his *consulti*, would be highly desirable in several respects: it would lead to a different perception of seventeenth-century medicine, clarify the status of those very anatomical discov-

7 Biblioteca Universitaria, Bologna, MS 936 I, esp. f. 25v. C. Dollo, *Modelli scientifici e filosofici nella Sicilia spagnola* (Napoli, 1984), pp. 154–5 n. 60. D. Bertoloni Meli, 'The posthumous dispute between Borelli and Malpighi, in Bertoloni Meli, *Malpighi* (n. 1). *Memorie di me Marcello Malpighi a i miei posteri fatte in villa l'anno 1689* (Bologna, 1902). M. Malpighi, *Opera Posthuma* (London, 1697).

8 Several such statements are reported by Malpighi's followers. See for example D.M. Schullian (ed.), *The Baglivi Correspondence from the Library of Sir William Osler* (Ithaca, 1974), p. 147: Baglivi to Manget, 1 April 1694.

9 M. Malpighi, *Consultationum medicinalium centuria prima* (Padua, 1713); G. Armillei, *Consulti medici di vari professori* (2 vols., Venice, 1743–5); M. Malpighi and G.M. Lancisi, *Consultationum medicarum editio novissima* (Venice, 1747). English translations of thirteen consultations by Malpighi from the 1747 edition can be found in S. Jarcho, *The Concept of Heart Failure from Avicenna to Albertini* (Cambridge, Mass., 1980), pp. 190–209. The volumes by Armillei, containing dozens of consultations by Malpighi, were not known to Adelmann.

eries by Malpighi which have so far attracted the greatest part of the historians' attention, and shed light on the fate of iatromechanics in the generation following Malpighi. It is worth recalling here that our strong preference for theory was not shared by his disciples in the eighteenth century, including Albertini, Antonio Maria Valsalva, Antonio Vallisnieri, Giorgio Baglivi, Lancisi and Giovanni Battista Morgagni.

Are Malpighi's *consulti* part of his *Correspondence?*

The specific aspect I have selected for this essay is the status of Malpighi's *consulti* in relation to his correspondence published by Howard Adelmann in 1975. This topic shows how the criteria adopted by an editor – an extraordinarily reliable one, to be sure – shape the image and our perception of historical actors and of their discipline. The problems encountered by Adelmann with Malpighi's *consulti* are best exemplified by a passage from his introduction to the correspondence:[10]

> *Consulti*, when unmistakably such, have not been included. A distinction between a *consulto* and a personal letter is, however, sometimes difficult to make. My practice has been to regard as a letter any communication provided with a personal salutation, even though in content the letter amounts essentially to a *consulto*. It is to be hoped that in the near future someone with the proper qualifications will edit and publish Malpighi's *consulti* with suitable medical annotations.

It is regrettable that Adelmann has been so modest, because the recent edition of Malpighi's *Consulti* falls short of his high standards by frequently omitting to indicate whether a text has already been published in The Correspondence of Marcello Malpighi, as well as in editorial practices and scholarly apparatus. In addition, there are other reasons to be unhappy with Adelmann's decision.

At the most straighforward level, it is not difficult to find *consulti* in the three manuscript volumes preserved at the Accademia delle Scienze dell'Istituto di Bologna studied by Adelmann, for example, which have not been included in the edition of the correspondence, although they contain personal salutations.[11]

A slightly more sophisticated survey of the sources throws further doubts on Adelmann's decisions. *Consulti* may survive through different channels. First, there are the original documents which were dispatched either to the patient or the attending physician. Second, there are drafts retained by the author for his own records. Third, there are copies generally made by other physicians either of the dispatched documents or of the ones in the author's possession. Many of Malpighi's *consulti* survive in copies, such as those in the three volumes at the Bologna *Accademia* in the hand of Giovanni Battista Morgagni. It is therefore legitimate to suspect that either Malpighi himself when he was copying *consulti* for his own

10 *Correspondence*, i, xiv.

11 See for example *Consulti*, i, 70–1 (. . . facendole umilissimo inchino, mi confermo per sempre etc.); ii, 132 (Excellentissime vir); iii, 64 (Excellentissimo et praeclarissimo viro domino Ioseph Maria Zamperio . . .).

records, or Morgagni, or an intermediate copyist, deliberately omitted forms of salutation contained in the dispatched documents and retained the strictly medical material, adding a suitable heading on the patient's disease. This was standard practice because epistolar *consilia* tended to become manuals where personal references, such as names of the patient or attending physician and forms of salutation, were frequently omitted and the texts arranged according to illness. Indeed, there are cases when the dispatched version and a copy of a *consulto* survive, showing evidence of these transformations. The letter by Bernardino Ramazzini to Malpighi of 4 September 1689 starts with the salutation: 'Illustrissimo Signore Signore Padrone Colendissimo', whereas in the copy in Morgagni's hand this is omitted and the *consulto* bears the title 'Difficilis Respiratio'. The salutation in Malpighi's reply to Ramazzini is also omitted.[12] This instance shows that, even if one were to adopt Adelmann's criterion, this should be applied only to dispatched documents or complete transcriptions, not copies intended to form a medical collection. This explains Adelmann's inclusion of *consulti* such as those written to Antonio Ferrarini and Bernardino Ramazzini at Modena and Angelo Modio at Rome, where dispatched documents can be traced, and the exclusion of others.

In this essay I shall argue an even more radical stance, namely that to my knowledge, all Malpighi's *consulti* are *ipso facto* epistolar exchanges and should therefore be part of his correspondence. Even if many *consulti* without forms of salutation exist among Malpighi's own papers, we have to suppose either that Malpighi omitted to transcribe them in the documents he retained for his own records, or that forms of salutation were enclosed with an accompanying sheet. In both cases the presence of forms of salutation must be considered as an accidental feature rather than as a classification criterion. I wish to qualify my statement by arguing that not necessarily all *consulti* by any author can be classed as letters. There are instances of texts called *Consilia* which were not written for a specific patient, but were rather general treatises on a disease and its therapy, or on several diseases and their therapies.[13] However, this is not our case.

What are Malpighi's *consulti*? They are written replies to a request regarding a specific patient, coming from either the patient himself or herself, a relative,[14] or, most frequently, a physician following their case. The request for a *consulto*, or letter of referral, usually came from a different town, thus Malpighi would normally comment on the case *in absentia*. This remained standard practice for

12 *Correspondence*, iv, 1504: Ramazzini to Malpighi (see also pp. 156 and 158); *Consulti*, i, 176 (see also pp. 178, 179). See also S. Jarcho (ed.), *Clinical Consultations and Letters by Ippolito Francesco Albertini, Francesco Torti and other Physicians* (Boston, 1989), pp. xix–xx. See also Dollo (n. 1), pp. 209–10.

13 J. Agrimi and C. Crisciani, *Les Consilia Medicaux* (Turnhout, 1994), pp. 19–21. For a general characterisation of *consilia* see D.P. Lockwood, *Ugo Benzi. Medieval Philosopher and Physician 1376–1439* (Chicago, 1951), esp. pp. 47–78, 78–138, and 323–34; N. Siraisi, *Taddeo Alderotti and his Pupils* (Princeton, 1981), ch. 9.

14 See for example *Consulti*, i, 171–3; ii, 182, 313, 376; iii, 118–19 (letters written by patients). The last letter, by Erminia Santa Croce Lancelotta, is extremely interesting because the patient challenges the therapies prescribed and reports contrasts among the attending physicians.

over a century after Malpighi's death, including a central figure in pathological anatomy such as Morgagni.[15] Unlike a case history, a *consulto* usually does not give any indication as to the outcome of the therapy, though occasionally this was later annotated in the manuscript with a brief explanation or a cross. The preceding reflections raise the interesting issue of how Malpighi and other physicians transcribed and organised their own and other practitioners' *consulti*. From the headings it appears that they were classified according to the disease or organ affected, rather than, for example, the name of the patient.[16] Interestingly, similar criteria were adopted by Malpighi in his manuscripts MS 2085, XII–XIII discussed above.

As a literary genre, the *consulti* date back to several centuries before the early modern era, were particularly common in Italy, and seem to share a common origin with legal documents. On the whole, many features remained constant over the centuries, though variations in style do occur depending on time of composition, local traditions, and personal style of the physician. By and large, Malpighi's *consulti* are highly structured and follow a common pattern, though I shall add some qualifications to this statement below. They are written both in Italian and Latin. As we have seen, they may begin with a form of salutation, followed at times by a brief summary of the case, including sex, age, constitution, and symptoms or *symptomata* of the patient. Since these aspects were provided in the letter of referral, there was no need for Malpighi to repeat them at the ouset. Indeed, he often started by praising the author of the letter of referral for his accurate account. Thus Malpighi could move directly to the second part, which is of considerable interest to historians of anatomy and medical historians concerned with the relations between theory and practice because there he provided an extensive causal account of the disease largely dependent on his anatomical findings. Probably the novelty of his anatomy was a factor prompting Malpighi to indulge in an extensive etiology. In the context of his explanation Malpighi frequently referred to the symptoms mentioned in the letter of referral as effects of the causes he had just provided. Typical accounts were based on the lack of fluidity of the blood, its excessive acidity or fermentation, and the malfunction of glands. Although I shall come back to the importance of this material in my final section, here I provide as an example the following explanation of bloody sputum in a noble patient, where Malpighi relies on his microscopic investigations on lungs:[17]

> For it is clear that active, sharp, and sylvestrine salts are present in gross excess in the noble patient's fluids, in which the mass of the humours is dissolved. Consequently, when the passages in the glands are relaxed, an

[15] On this important aspect see W.F. Bynum, 'Health, disease and medical care', in G. Rousseau and R. Porter (eds.), *The Ferment of Knowledge* (Cambridge, 1980), pp. 211–53, on pp. 212–13; G.B. Morgagni, *Clinical Consultations. The Edition of Enrico Benassi (1935), translated and revised by S. Jarcho* (Boston, 1984), p. lii.

[16] See Biblioteca Universitaria, Bologna, MS 2085, IV–V and XIV–XVI.

[17] S. Jarcho, *The Concept of Heart Failure* (n. 9), p. 202, transl. from Malpighi and Lancisi, *Consultationum medicarum editio novissima* (n. 9), pp. 43–4. I have altered Jarcho's interpolations in square brackets.

abundance of abnormal saliva bursts forth. This is made so salty and annoying by the admixed erosive particles that the gums are eroded as if by *aqua fortis* and the teeth become carious. Befouled by the same particles as the blood approaches the passageways in the lungs, it gradually produces congestion, for particles not thoroughly mixed in are intercepted, and, when they become lumped together, they impede the circulation until under the impulsion of the rest of the blood the delicate little membranes in the lungs are taken apart and the blood bursts forth into the pulmonary vesicles and the ends of the trachea, and shows itself as bloody sputum.

The causal account of the disease was followed by a brief set of *indicationes* or indications, usually laid down in two or three sentences. These were lines of action expressing the general aim of the therapy, such as render the blood more fluid or less acid. A typical example is the following:[18] 'Hence, the indications are: to remove the abundant, acid, erosive particles, to restore tonic movement in the solid structures, and to strengthen the ferments in the viscera.'

The *indicationes* were followed by the therapy based on *remedia* or remedies, usually an extensive list of medicaments, suggestions on how to prepare them, and length of time necessary for the cure to have effect. In one case Malpighi even suggested the chemists where certain pills were sold, namely the 'Spezieria del Sole' in the Galliera quarter. The 'Spezieria del Sole' belonged to Angelo Michele Cantoni, the best man at Malpighi's wedding in 1667 who was still considered in 1691 one of his two best friends. Several letters and *consulti* refer to Cantoni's ways of preparing drugs.[19] Before the final form of salutation and the prayer 'Faxit Deus' or 'May God act favourably', Malpighi frequently stated that he was submitting his *consulto* to the better judgement of the physician *in loco*.[20]

It should be emphasised that this is the structure of a representative *consulto*, though several were much shorter and lacked entire sections. Unfortunately, at present our knowledge of Malpighi's *consulti* is so limited that we do not know whether his therapies changed with time and anatomical discoveries, whether there are differences between Latin and Italian texts, whether the profession of the addressee affected his style of writing, and how his medical practice relates to that of other physicians.[21]

18 S. Jarcho, *The Concept of Heart Failure* (n. 9), p. 195, transl. from Malpighi and Lancisi, *Consultationum medicarum editio novissima* (n. 9), pp. 6–7.

19 Belloni, *Opere* (n. 3), p. 471: 'Caso poi, preso il decotto, che restasse qualche vestigio di gonorrea, prender certe pillole che si dispensano in Bologna alla Spezieria del Sole in Galliera'. Adelmann, *Embryology* (n. 1), p. 320. *Correspondence*, iv, 1678 and ii, 685, iii, 1002, 1157, 1259, 1338, iv, 1531, 1565, 1567, 1635, 1817, 1830, v, 1915.

20 This caveat was standard in Morgagni's *consulti*. From a preliminary survey it appears that Morgagni devoted more space to summarising the history of the case and the patient's symptoms, and displayed his erudition by quoting extensively from the medical literature, a practice followed more sparingly by Malpighi.

21 A fascinating letter on venesection by Giovanni Battista Capucci emphasised the differences of opinions in the neoteric camp: *Correspondence*, i, 336–8. Capucci reported that Tommaso Cornelio at Naples was opposed to venesection, whereas he performed it only in order to please his patients. Se also L.A. Porzio, *Erasistratus, sive de sanguinis missione* (Rome, 1682).

Contrasts on the opportunity to publish Malpighi's *consulti*, especially those perceived to be extemporaneous observations rather than extensive and well-pondered texts, emerged among the first generation of Malpighi's followers. When Antonio Vallisnieri's pupil, Girolamo Gaspari, published a *Centuria* of Malpighi's *Consulti* in 1713, Ippolito Francesco Albertini was outraged because Malpighi's own wishes had been violated and because Gaspari's edition was fraught with errors. He even issued a *Monitum Literarium* listing the most significant mistakes, such as 'lacte et vino' instead of 'lacte vaccino', or 'fermentatione' instead of 'ferruminatione', which he had checked against the documents in his possession. The correspondence between Vallisnieri and Albertini on these events highlights different perspectives: the latter complained that Gaspari had published many *privati biglietti* or brief notes which did no honour to Malpighi. The former argued that precisely those brief accounts often contained profound insights and compared them to the sketches of a great painter.[22] Such differences of opinion are potentially extremely important in explaining which texts have been preserved and which have been excluded by a physician interested in compiling a collection of Malpighi's *consulti* for his own practice. This reflection should sound as a warning not to take the three volumes collected by Morgagni at face value, or as a representative selection of all the available material. Morgagni's taste and preferences almost certainly played an important role in the way he selected the manuscripts of his mentors' teacher.

Malpighi's *Consulti*: from mechanistic anatomy to mechanistic medicine

The consequences of Adelmann's decision to exclude Malpighi's *consulti* in his edition of the correspondence are considerable in several respects. On a quantitative level, the inclusion of all the letters of referral and *consulti* in Morgagni's hand recently published would add two volumes to the five produced by Adelmann. This, however, is a conservative estimate, since other manuscript volumes of *consulti* exist, as well as smaller collections and single items. This material still awaits systematic investigation, therefore a more reliable estimate of how many volumes could have been added to Malpighi's correspondence cannot be given at present.

Despite this problem, it is not difficult to grasp the import of these additional documents. They cover a large portion of Malpighi's life and testify to his daily practice even at times when no other correspondence survives. Moreover, the geographical provenance of the letters of referral and the social conditions of his patients give us a vivid picture of Malpighi's sphere of action and status as a physician.

22 Atti, *Notizie* (n. 1), pp. 421–47. Atti provides an extensive reconstruction of the controversy, including several unpublished letters and Albertini's *Monitum Literarium*.

Besides these social reasons, Malpighi's *consulti* have an extraordinary intellectual import in showing how mechanistic medicine worked at the bedside. The causal and anatomical sections of his *consulti* provide links between his views on the animal oeconomy – notably his findings on the lungs, glands, heart polyp and blood – and the notion of disease. Although in his anatomical writings Malpighi privileged iatromechanics, in his *consulti* he established closer links with iatrochemistry: the malfunction of glands was often attributed to a defective ferment affecting their ability to filter the proper juices. Therefore, the chemical composition of those juices, as well as of blood from which they were believed to be filtered, was altered and had to be corrected by therapy.[23]

Traditional accounts of the scientific revolution dealing with mechanistic anatomy emphasise Descartes's role as philosophical mentor of the iatromechanical school, Malpighi's virtuosity with the microscope and discovery of the blood capillaries in the lungs of frogs in *Epistolae de pulmonibus* (Bologna, 1661), culminating with Giovanni Alfonso Borelli's masterpiece *De motu animalium* (Rome, 1680–81), where quantification took a central role. From a survey of the late seventeenth-century European scene, however, it appears that Malpighi's *De viscerum structura* (Bologna, 1666) represented not only one of the major anatomical achievements of the iatromechanical school, but also a crucial text for establishing a link between anatomy and the causes of disease.

Unfortunately, how mechanistic medicine worked in practice is not part of the general picture. One of the reasons why this area has been frequently neglected is that several historians deny a role to medical practice in the intellectual revolution of the seventeenth century, though this can hardly be a satisfactory justification.[24] Malpighi's claims that the liver, kidneys, cerebral cortex and, to some extent, the spleen consist of glands, for example, represented a major aspect of his new medical framework. If we want to reconstruct the social and intellectual horizons of iatromechanical physicians, we have to pay attention to medical practice and the *consulti* must become an integral part of our picture, rather than an embarrassing appendix which is best left to technical medical historians. This is an area where we have to reconcile the heroic image of Malpighi as the founder of microscopic anatomy with the *remedia* he prescribed: ground

[23] *Consulti*, ii, 17 on the mulfunction of the glands in the brain. Malpighi believed to have discovered them in 'De cerebri cortice', published in *De viscerum structura* (Bologna, 1666). E. Clarke and J.G. Bearn, 'The brain glands of Malpighi elucidated by practical history', *Journal of the History of Medicine*, 23 (1968), 309–30.

[24] On this topic see H.J. Cook, 'The new philosophy and medicine in seventeenth-century England', in D.C. Lindberg and R.S. Westman (eds.), *Reappraisals of the Scientific Revolution* (Cambridge, 1990), pp. 397–436; A. Wear, 'Medical practice in late seventeenth- and early eighteenth-century England: continuity and union', in R. French and A. Wear (eds.), *The Medical Revolution of the Seventeenth Century* (Cambridge, 1989), pp. 294–320; A. Guerrini, 'The varieties of mechanical medicine: Borelli, Bellini, Malpighi, and Pitcairne', in Bertoloni Meli, *Malpighi* (n. 1). R. Porter, 'Strategie terapeutiche', in M.D. Grmek, *Storia del pensiero medico occidentale. 2. Dal Rinascimento all'Illuminismo* (Roma-Bari, 1996), pp. 335–80.

human skull against epilepsy,[25] viper's meat,[26] crab's eyes, hartshorn jelly,[27] spirit of human blood,[28] and decoctions with children's urine[29] are just a few of the ingredients recommended by Malpighi.[30] Even if several of these remedies can be documented as dating back to antiquity, the framework justifying their use changed with time, thus the very same remedy could be employed for entirely different reasons. Therefore they should be seen not in isolation, but rather in conjunction with the theoretical and experimental practices associated with them. In this context Malpighi's close friendship with the apothecary Angelo Michele Cantoni appears to be a matter of considerable interest because our sources suggest that personal friendship went hand in hand with some degree of intellectual co-operation.[31]

In conclusion, I would like to instantiate my claims about Malpighi's *consulti* with an example emphasising the significance of this neglected genre not only from a social, but also from an intellectual standpoint. In a *consulto* for a Milanese signore, dated 18 June 1692 and later titled 'Morbo Melancholico', Malpighi argued that the most noble patient was affected by a delirium without fever which made him believe to be affected by many non-existent diseases. In order to account for this delirium, Malpighi indulged in a series of *congetture* on how the brain and phantasy work based on a combination of Cartesian views and his own anatomical findings, especially those in his 1665 *De cerebro* and 1666 *De cerebri cortice*, where he believed he had discovered glands in the cerebral cortex separating the nervous juice:[32]

> The structure of the brain seems to consist in a network of fibres, or small cavities filled with a juice which is separated in the cerebral cortex, whence they originate. They are moved by the external organs and further by the external objects through the nervous connections, and receive a trembling motion which is transmitted in an ineffable fashion to the soul and excites

25 This remedy, in conjunction with others, was commonly prescribed: see for example *Consulti*, ii, 202 and iii, 12. Human skulls were also recommended by Morgagni. I wish to emphasise that any seventeenth-century definition of a disease, such as epilepsy, does not necessarily correspond to the modern notion bearing the same name.

26 *Consulti*, i, 111. On the use of vipers as late as the second half of the eighteenth century see the essay by E. Benassi, 'Morgagni's use of vipers' in Morgagni, *Clinical Consultations* (n. 15), pp. 316–18.

27 These two remedies are among the most frequently recommended by Malpighi.

28 *Consulti*, iii, 33.

29 *Consulti*, iii, 44–5.

30 Standard medicinal collections were listed in the *Antidotarii* issued in several cities. See for example *Antidotarium Collegii Medicorum Bononiensis* (Bologna, 1770).

31 For an instance regarding the use of milk for curing fevers see D. Bertoloni Meli, 'The new anatomy of Marcello Malpighi', in Bertoloni Meli, *Malpighi* (n. 1), section 3. On a recent debate on the modernity of the *Consulti* by Redi see F. Redi, *Consulti medici*, ed. C. Doni (Florence, 1985, *Politica e storia. Saggi e testi*, 9); B. Basile, *Linvenzione del vero* (Rome, 1987), chapter 3, and his review of Doni in *Studi e problemi di critica testuale* 32 (1986), 197–200. Basile claims that Redi was post-Galilean with regard to theory and pre-Galilean in medical practice.

32 *Consulti*, i, 244–5 (also published in the 1747 edition of *Consulti* by Malpighi and Lancisi):
La struttura del cervello pare che sia un plesso di fibre, o fistole ripiene d'un sugo, che si separa nella corteccia, dove sono radicate. Queste mosse mediante la continuazione de' nervi dagli organi esterni e successivamente dagli oggetti estranei ricevono un tremore che

> passions in it. Since the same movements can be communicated by the motion or quality of the nervous juice, or by other bodies touching those fibres, without external objects, it follows that without external motion the soul can feel and those very passions which external objects produce through the senses can be excited in us.

Malpighi argued that the plausible cause of his patient's illness was an excess of acid juices preventing the cerebral fibres from having the right tension. That excess was inferred from a quartan fever two years previously, a sedentary life whereby acid juices were not eliminated and ended up in the brain, and possibly the corrupt glands of the lower belly and especially the pancreas. The *indicazioni* are to remove the excessive acid juices in the brain, clean the juices of the lower belly, make the fluids more volatile and favour transpiration. The long list of *rimedi* includes venesection, herbal juices, baths in fresh water in order to increase transpiration, strained whey of goat's milk, crab's eyes, local motion, conversation, and change of air.

This example suggests a reason why Malpighi was reluctant to have his *consulti* published, namely that in those private documents he had indulged in a series of conjectures and speculations which he would not have wished to commit to the press. Despite their different status from the *Epistolae de pulmonibus*, however, his *consulti* 'throw light on individual problems that confronted a distinguished physician and set forth his reasoning, his real or professed opinions, and his therapeutic resourcefulness'.[33] For these reasons they ought to be an integral part of our image of Malpighi and mechanistic medicine.

Besides focusing on the relations between *consulti* and correspondence, the preceding reflections highlight a more general aspect. A correspondence, by its very nature, does not include exchanges with that restricted group of friends whom the author may have met in his town on a daily basis. This is the reason why only one letter between Malpighi and one of his two closest Bolognese friends is extant. The fact the Angelo Michele Cantoni was an apothecary deserves greater attention by medical historians and points, jointly with the issue of the *consulti*, to the need for a profound reassessment of the links between Malpighi the anatomist and Malpighi the physician. The examples provided in this essay instantiate the more general claim that a fresh and critical look at an author's archive reveals an exciting and problematic picture which we are just beginning to investigate.

> communicato con un modo ineffabile all'anima, eccita in essa le passioni. E perchè gli stessi muovimenti si possono communicare senza gli oggetti esterni dal moto, o qualità del sugo nerveo, o vero da altri corpi che toccano dette fibre, quindi ne nasce, che senza mozione esterna l'anima può sentire ed in noi si possono eccitare quelle passioni, che producono mediante i sensi esterni gli oggetti fuori di noi.

My account of this *consulto* aims at giving an outline of Malpighi's reasoning. The reader interested in a detailed analysis of medical matters is advised to read the original. R. Descartes, *Les passions de l'âme* (Paris, 1649), esp. articles 26–7. See also T. Willis, *Cerebri anatome; cui accessit nervorum descriptio et usus* (London, 1664) and the tercentenary edition by W. Feindel, *The anatomy of the brain and nerves* (2 vols., Montreal, 1965).

33 Jarcho, *The Concept of Heart Failure* (n. 9), p. 190.

8

Mapping the Mind of Robert Boyle: the Evidence of the Boyle Papers

Michael Hunter

The Boyle Papers and Letters, which have been at the Royal Society since 1769, represent a body of material of a size comparable to any of those dealt with in this volume. They comprise seven guardbooks of letters, forty-six volumes of papers, and some twenty notebooks and other discrete manuscripts.[1] In total, they comprise some 15,000 folios. Much of their content represents the personal manuscript remains of Robert Boyle – his notebooks, drafts, records of experiments, letters and memoranda. But mixed in with this is a great deal of more miscellaneous material: translations of Boyle's writings into Latin; copies of treatises that interested him on alchemy, medicine, travel and related subjects; and copies of political and religious tracts relating to the issues of the day, at least some of which may have come into the archive in the generation after his death.[2]

The material is bound in volumes, but the main thing that can be said about its current ordering is that it is certainly not Boyle's. The letters are in an alphabetical sequence, while the Papers are very roughly arranged into the categories of 'Science', 'Philosophy', 'Theology', 'Physiology' and 'Miscellaneous'. These were imposed on the archive in the 1850s, when the manuscripts were bound in their present form. On the other hand, the sorting then executed was haphazard in the extreme. In many cases, parts of the same treatise are split between two different volumes, while, even when items are all together, they are often bound in the wrong order.

Before the 1850s, we know of at least two campaigns of work on the archive. One took place around 1740 in conjunction with the preparation of Thomas Birch's edition of Boyle's *Works* in 1744, which, in its reprinted version of 1772, remains the standard edition, pending the appearance of a new one within the next few years. The relevant archival work was mainly done by Birch's collaborator, the

I am grateful to Edward B. Davis for his comments on a draft of this paper.

1 See Michael Hunter, *Letters and Papers of Robert Boyle: a Guide to the Manuscripts and Microfilm* (Bethesda, Md., 1992), and the important supplementary study by Lawrence M. Principe, 'Newly Discovered Boyle Documents in the Royal Society Archive: Alchemical Tracts and his Student Notebook', *Notes and Records of the Royal Society*, 49 (1995), 57–70.

2 For the points in this and the next three paragraphs, see Hunter, *Letters and Papers*, esp. pp. xvii–xxviii.

dissenting minister, Henry Miles, in whose hands the Papers then were, to whose work in sorting and rearranging material various endorsements, transcripts, lists and memoranda testify.

Earlier still, extensive work was done on the archive by the scholar William Wotton, who carried out much research in the years around 1700 for a life of Boyle, though this never materialised. Wotton's activities are attested mainly by profuse endorsements, illustrating his attempt to impose a system of classification by files and bundles. The exact nature of Wotton's classification is unclear, since hardly any of his own lists and notes survive, but it does not seem as if he was ratifying Boyle's own arrangement of the papers so much as imposing a new one. He thus began a process of eradicating the traces of Boyle's own disposition of the papers through well-intentioned efforts to 'organise' the archive which has continued ever since.

If it is clear that the archive has been rearranged since Boyle's time, it is equally clear that it has suffered attrition, particularly at some point between Miles' work on it and its presentation to the Royal Society by his widow in 1769, though there was almost certainly also an element of loss between Wotton and Miles. The scale – if not the full extent – of the loss is clear from Miles' inventories, which include many items that are no longer extant. Some of the losses seem fairly random, but others were not, and they have had a distorting effect on the archive which needs to be taken into account in interpreting it. In particular, there seems to have been a tendency to discard papers which seemed trivial or peripheral to Boyle's intellectual concerns, which has the result of making seem Boyle more Olympian and detached from his context than would otherwise be the case. There was also an attempt to obscure his links with the highly-charged religiosity of his day, and to play down his involvement in alchemy and other pursuits that smacked of credulity.[3]

Here, I want to try to get beyond both such losses and the various reorganisations that the papers have suffered to probe at the form that the archive took during Boyle's lifetime – and in particular the way in which he arranged it and deployed it. This might seem an elusive topic, particularly after the rearrangements to which the archive has been subjected over the intervening period. Indeed, these have inspired various twentieth-century commentators on the archive, such as Marie Boas Hall and M.A. Stewart, to complained of its 'disarray'. In this, they echo Miles, who was inclined to blame the problem on Wotton.[4] But what has not generally been recognised is that the perception of the archive as chaotic can be traced back to Wotton himself, within less than a decade of Boyle's death. Thus

3 See Michael Hunter, 'Robert Boyle and the Dilemma of Biography in the Age of the Scientific Revolution', in Michael Shortland and Richard Yeo (eds.), *Telling Lives in Science: Essays on Scientific Biography* (Cambridge, 1996), pp. 115–37, on pp. 128f. For an annotated text of Miles' inventories, see Michael Hunter and Lawrence Principe, 'The Lost Papers of Robert Boyle' (forthcoming).

4 M.B. Hall, 'Henry Miles, F.R.S. (1698–1763) and Thomas Birch, F.R.S. (1705–66)', *Notes and Records of the Royal Society*, 18 (1963), 39–44, on p. 42; M.A. Stewart, 'The Authenticity of Robert Boyle's Anonymous Writings on Reason', *Bodleian Library Quarterly*, 10 (1978–82), 280–9, on p. 283. For Miles' views, see Hunter, *Letters and Papers* (n. 1), pp. xxv–vi.

in 1699, soon after Wotton first gained access to Boyle's manuscript remains, he replied to John Evelyn's request for material 'which may relate to the res rustica' by remarking: 'His Papers were truly, what he calls many Bundles of them himself a Chaos, rude & indigested many times God know's.'[5] It did not help that he was apparently sent the papers in sections, rather than being given the opportunity to peruse them as a whole. But this may suggest that, like all subsequent commentators, Wotton was looking for a kind of order that did not really exist, and the apparent 'disarray' of which later scholars have complained may have been due less to depredations suffered by the archive after Boyle's death than to his own way of keeping it. Indeed, there are certain clues which suggest that the chaos may have been more apparent than real, and that in fact the archive had an ordering which bore a significant relationship to Boyle's intellectual aims and methods. It is these that I wish to explore in this paper.

Relatively little evidence survives in the form of Boyle's own cover-sheets and other such physical remnants of his organisation of his papers.[6] On the other hand, what we do have is a series of lists and inventories made at Boyle's behest, particularly in his later years, many of them dated, and many of them providing clues to the physical state of the manuscripts.[7] These provide crucial evidence concerning the form that the archive took during Boyle's lifetime, and I want to itemise them and indicate their significance here. First, there is a pair of lists in the hand of Henry Oldenburg. One is entitled 'A List of such papers, as were mentioned to me by the Honourable Robert Boyle March 26 1677', and the sequel similarly itemises 'such papers of Divinity' as Boyle told Oldenburg about on the same day.[8] The first list includes the titles of various works then impending, including *The Producibleness of Chymical Principles*, the sequel to the *Sceptical Chymist* which came out in 1680. It also has a variety of more miscellaneous items, including 'A Book of Schemes and Queries relating to Natural philosophy', and 'A Chaos of Experiments of my owne, consisting of nine Centuries. In three or four several paper books'. The second list similarly itemises works by Boyle which were subsequently to appear in print in the next few years, this time religious, such as his *Discourse of Things above Reason* (1681), together with other more miscellaneous items. In the margins to both lists (though more systematically in the first than the second), Oldenburg has added a note of the format of each item – for example, 'In an orange-browne paper-cover' or 'In a blew cover with labels of cards'. We will return to the significance of these shortly.

A further, undated list may be assigned to c. 1680 on the evidence of handwrit-

5 Wotton to Evelyn, 8 Aug. 1699, British Library Evelyn Collection MS 3.3.112. For Evelyn's request, see ibid, 110–11.

6 There are, however, a few examples: see Royal Society Boyle Papers (hereafter BP) 8, fol. 63 (see further below), or the original cover bound into Royal Society Manuscript (hereafter MS) 200.

7 For earlier lists, which give no evidence on how the manuscripts were stored, see BP 36, fols. 85, 86, 70 and 177, and below. An edited text of all of these inventories will appear in the forthcoming 'Pickering Masters' edition of *The Works of Robert Boyle*, ed. Michael Hunter and Edward B. Davis, vol. 14.

8 BP 36, fols. 88–9.

ing.[9] It itemises the contents of 'the Box A' and 'the Box B', listing various volumes or bundles of 'Memorandums', 'Communicated Observations', 'designed Experiments' and experimental notes, some divided into 'centuries' and including 'the Nuington Diary'. There are also volumes comprising 'Reason Bowing to Religion', 'Promiscuous Papers about Tobaco the Aire &c.', while more specific items include 'the re[quisites] of a Good Hypothesis & the improvement of Materia medica' and 'A Purple Paper about humane Bloud &c.', the latter evidently linked to Boyle's medical publications in the 1680s.

The next group of lists dates from July 1684, and is in the hand of the amanuensis whom Boyle most commonly employed in his later years, Robin Bacon. First, we have 'A Catalogue of the Honourable Mr Boyles Writings unpublish'd taken the 7th of July 1684'.[10] This sounds as if it is just going to be a list of titles, but it also includes what were clearly bundles of papers, including collections of medicines and of extracts from printed books; there are also such items as 'A Rapsody of rude schemes of Discourses of Naturall History and other Subjects'. Linked to this is a further document, dated as being compiled three days later, which itemises 'Papers left in the Box A' and 'In the Box B'.[11] These might have been the same boxes as in the undated list (though Box A is here identified as being of 'Marble Colour'), but the content is by this time quite different. Box A now contained theological treatises, including writings on the mutual relations of reason and religion, together with Book 2 of Boyle's early religious romance, *The Martyrdom of Theodora, and of Didymus*, which he rewrote at this time and published in 1687, and more miscellaneous items. Box B had miscellaneous scientific material, some of it specified as being on such subjects as 'Light & Inflammability', or 'the Generation & Transmutation of Metall'. Again, the format and appearance of certain of the items in question is indicated in the margin.

The next two lists appear in the notebooks which Boyle kept in his later years and which survive as part of the archive.[12] The first is simply an undated list of Boyle's published and unpublished theological writings, but the second, dated 22 July 1688 and entitled 'A list of Papers kept in the lower part of the new Press standing between the scrittore & the great Bed', itemises a mixture of material stored there. Many of these were writings by Boyle, mainly scientific though including one theological item, but there were a couple of extraneous items – 'A Relation of [the] last new English war with the Indians 4to' and the preface to the edition of the New Testament in Irish with the publication of which Boyle was associated in the 1680s.[13]

Then, there are various lists dating from 1691, the last year of Boyle's life. The

9 BP 36, fol. 61. This is in hand C, on which see Michael Hunter and Edward B. Davis, 'The Making of Robert Boyle's *Free Enquiry into the Vulgarly Receiv'd Notion of Nature* (1686)', *Early Science and Medicine*, 1 (1996), 204–71, on pp. 212–13 n. 21.

10 BP 36, fols. 59–60, a fair copy of ibid., fols. 75–6.

11 BP 36, fols. 119–20. For other copies, see ibid., fols. 117–18; BP 35, fols. 192–3.

12 MS 185, fols. 1, 3; MS 191, fols. 1–2. On the notebooks, see further below.

13 See R.E.W. Maddison, 'Robert Boyle and the Irish Bible', *Bulletin of the John Rylands Library*, 41 (1958), 81–101.

first, again to be found in one of Boyle's notebooks and dated 1 January, lists the 'Papers in the long Box', and comprises a miscellany, including 'the Latin version of the Appendix to the Sceptical Chymist' and such items as an account of Virginia addressed to Boyle.[14] A further list, dated 7 January 1691, has three headings, each referring to a box and giving the instruction: 'seek for these writings'.[15] (One of the boxes is described as 'Marble Paper'd', and this is perhaps that called Box A in the 1684 list.) The fact that this was a list of items to be looked for means that this is presumably not a complete inventory of the content of each box; it is perhaps not surprising that a disproportionate number of the items relate to specific works, notably Boyle's *The Christian Virtuoso* and his *Experimenta et Observationes Physicae*, the former recently but incompletely published, the latter published in the summer of that year so perhaps already in the press, and it may have been associated with a re-ordering of his papers on Boyle's part. Then, another list dated 21 January without year, but clearly belonging to 1691, has a list of the contents of five boxes, each described as 'markt . . .' but without the mark being divulged.[16] Two of these have contents similar to the items listed as to be searched for in the list of three boxes of two weeks earlier – one combining *The Christian Virtuoso* and medical items, the other devoted to *Experimenta et Observationes Physicae* – but the contents of the third are now split into two, each mixed with other material in a discrete box, while there is also a fifth box containing more miscellaneous theological items.

The final group date from the summer of 1691. First, there is a list of Boyle's philosophical and theological writings, 'not yet printed, set down July the 3rd 1691', which predominantly lists titles, though it includes some more miscellaneous items such as 'A Chaos Chymico-medical'.[17] The final inventory is dated 17 September 1691.[18] This lists the contents of 'a Square Box in Mr Boyle's Bed-chamber', 'another Square Box mark'd . . . in his great Room', 'a green Bag in the same Room', and 'the long black Box (lockt) upon the Beds feet in Mr Boyls great Room'. Then, 'In Mr Boyles bed-chamber', it itemises sixteen numbered items 'In the upper Shelf of the Press in Mr Boyls Bed-Chamber', together with another twelve in 'a long Box lockt up under several shelves of Books'; the latter is clearly the same 'long Box' as was inventorised on 1 January, with overlapping, though not identical, content. Some of the material listed overlaps with what had appeared in earlier inventories, including boxes of material relating to *The Christian Virtuoso*, *Experimenta et Observationes Physicae*, and Boyle's medical writings, together with 'a thin Folio, containing many Prefaces Advertisements &

14 MS 186, fol. 37v. This will be dealt with in the forthcoming *Works* edition s.v. the inventory of 17 September 1691, with which it partly overlaps.

15 BP 36, fol. 73.

16 BP 36, fol. 121. Drafts survive in MS 186, fols. 36v–7, 38–9. In addition to the changes noted in the text, it is perhaps worth registering that the putative second part of *Medicina Hydrostatica* changed boxes.

17 BP 36, fol. 72. A duplicate survives in MS 35, fol. 187. For drafts, see MS 186, fols. 33v–4, 86v–7; see also BP 36, fol. 69.

18 BP 36, fols. 122–3.

such like Papers of great use for understanding the scope & method of many of Mr Boyl's unpublish'd writings'. But, in addition, there were Latin translations of Boyle's works, notes from books, material relating to treatises not previously noted in the inventories (such as Boyle's *Free Enquiry into the Vulgarly Receiv'd Notion of Nature*, published in 1686), volumes of correspondence, and a number of manuscripts of works by others: for example, 'Tryal of an Irish Witch' and 'A Bundle of Manuscript belonging to a french Minister Mr Boyle thinks Monsr. de Lortie'.

What can we learn from these lists? As should be apparent from their titles, their purposes are not identical: in particular, some are lists of unpublished treatises while others are simply inventories, and this explains some of the discrepancies between them. In addition, it is clear that none of the inventories is a complete listing of all Boyle's papers: quite apart from the chronological variation between them, different lists inventorise different sections of the archive, probably for different purposes. Hence care needs to be exercised in interpreting them, which was not shown by Henry Miles when he conflated various of them together to produce a composite and highly unsatisfactory list of unpublished works by Boyle which was included in Birch's edition and has often been cited.[19] But various conclusions can nevertheless be drawn, from both the similarities and the differences between the various listings. Moreover, that the bulk of the inventories deal with the core of the collection is suggested by the quite miscellaneous items that appear in the lists dated 1 January and 17 September 1691, which seem to deal with material secondary to this core, such as Latin translations of Boyle's works and copies of treatises by others.

In the first place, the inventories provide evidence of the method by which individual groups of papers were kept together. Indeed, they give a vivid insight into the range of eye-catching items of stationery made available by the burgeoning consumer economy of late seventeenth-century England and into Boyle's taste in it. Thus you could get 'cases' – presumably like box files – made of paste-board, i.e., board made by compressing and rolling together three or more sheets of paper, which came either in fine or coarse weights. You could also get folders made of paste-board with strings or ribbons embedded in them which could be used to tie them up, presumably like art students use today. These came in different colours, as did their attached strings/ribbons, which was a useful way of distinguishing different items; a further distinctive material used in covering folders and the like was marbled paper. Many of the characteristic identifications of materials in the inventories specify the details of the receptacles in which they were contained. For instance, the lists itemise material 'In [a] white past-board cover, whose strings are striped with white, dark and red', or 'In a white pastboard cover with parti-colored ribands', or 'In a Marble-paper Cover', or 'Quarto, Gray Past-board, black & green strings'.[20] It is perhaps of interest that one surviving item started

[19] Thomas Birch (ed.), *The Works of Robert Boyle*, 2nd ed. (6 vols., London, 1772), i, ccxxxvi–viii. For Miles' draft, see BP 36, fol. 155.

[20] BP 36, fols. 88–9, 119–20.

life as a grey wrapper with the title, 'Promiscuous Addenda to my severall Treatises' written on it, then had patterned paper stuck over it, most of which was subsequently torn off again.[21] Clearly Boyle found it helpful to retain a mental record of what notes were where in this way. Indeed, it is not fanciful to suggest that the folders provided a kind of colour-coded memory system, perhaps analogous to the suggestions as to how the memory might be aided in the process of data-retrieval made by Boyle's former protegé, Robert Hooke, in his 'General Scheme, or Idea Of the Present State of Natural Philosophy, and How its Defects may be Remedied' of c. 1668.[22]

The materials housed in these folders seem predominantly to have been unbound sheets, and I will return to the significance of this in a moment. But the inventories also refer to 'Paper Book[s] in folio' or quarto, in other words series of leaves stitched together, perhaps with covers, while, as an alternative to placing material in folders, it might just be tied up in 'bundles', 'parcels' or 'rolls'. For this, string or ribbon was used, which might again be distinctively coloured, and of which Boyle must have had a separate supply: thus we have items like 'A yellow 4to tyed up with Packthred writt upon Medicinalia' (pack-thread is a sort of twine used for tying up bundles).[23] A further piece of this 'literary technology', not directly referred to in the inventories but evidenced by passing references on Boyle's part, was the use of pins to pin together loose pages or fragments of less than a page in length, another characteristic consumer product of the time.[24]

Turning to the larger receptacles in which the collection was housed, the core of it appears to have been contained in a finite series of boxes, starting with Box A and Box B in the earliest inventories, and then multiplying in number in the later ones. In no list does the total number of these boxes exceed five, at least two of which were square, and connections can be made between boxes appearing in different lists, particularly the (evidently square) box covered with marbled paper which apparently began life as Box A. From this, it is apparent that the content even of the same box changed at different dates, and it seems likely that Boyle shifted into these boxes material that he perceived to be directly relevant to the treatises that he was working on at any particular point, though one item that recurs in various lists is a folder of 'Advertisements and Introductory Papers' which appear to have been kept together pending their transferral to the specific tracts or writings which they referred and to which an introductory note survives in one of Boyle's notebooks.[25] Otherwise, the shifting content appears to have related to Boyle's programme at the time. Thus the preponderance of items published in the early to mid-1680s in the lists of 1677 has already been noted, whereas the lists of c. 1680 and 1684 have a disproportionate amount of material relating to Boyle's

21 BP 8, fol. 63.

22 Hooke, *Posthumous Works*, ed. Richard Waller (London, 1705), pp. 63–5.

23 BP 36, fols. 122–3.

24 See, e.g., MS 189, fol. 121; MS 199, fol. 143. Cf. Joan Thirsk, *Economic Policy and Projects* (Oxford, 1978), pp. 78–83; John Aubrey, *Brief Lives*, ed. Andrew Clark (2 vols., Oxford, 1898), ii, 324.

25 MS 187, fol. 164v. Cf. the inventories of 1677, 1684 and 17 September 1691.

medical writings, many of which were published in the 1680s. By 1691, however, as we have seen, the predominant material related to *The Christian Virtuoso* and *Experimenta et Observationes Physicae*, to which Boyle had sequels in preparation at the time of his death. The way in which Boyle shifted material around is further suggested by the reference in the 7 January 1691 list to material to be searched for, while it is actually possible to observe identifiable folders being shifted between boxes in the sequence of inventories that survive from 1691.[26]

If the core of the collection appears to have been housed in a relatively small number of boxes, other listings describe what appear to have been more tangential items. Some of these – like the boxes – appear to have been housed in Boyle's principal rooms in the house in Pall Mall that he shared in his later years with his sister, Katherine, Lady Ranelagh, namely his 'Bed-Chamber' and his 'great Room' (which also seems to have had a bed in it). Indeed, the inventories give some sense of the physical conditions in which Boyle lived, bearing out John Evelyn's famous description of how his bed chamber was 'so extreamely crowded with Boxes, Glasses, Potts, Chymical & Mathematical Instruments; Bookes & Bundles of Papers; that there was but just roome for a few Chaires; so as his whole furniture was very Philosophical, without formality'.[27] On the other hand, a touch of class is added by the reference in the 1688 inventory to 'the new Press standing between the scrittore & the great Bed': this is in fact an relatively early instance of the use of the word 'escritoire', with its fashionable French connotations, to describe a writing-desk.[28] It is intriguing to speculate whether the 'Press' may have been an equally modish piece, perhaps similar to those which Samuel Pepys acquired to house his library and collections, which still survive in Magdalene College, Cambridge.

If some of Boyle's more tangential holdings were housed in this 'Press', or on shelves in his rooms, others must have been stored elsewhere in the house. One location where some were kept was in Boyle's laboratory, which appears to have been annexed to the house at the back.[29] Indeed, testimony to one manuscript's being in the laboratory is provided by the fact that an opening within it has clearly had a chemical substance spilt all over it; adjacent to this are a series of finger and palm prints evidently linked to mopping it up. This curious document would repay forensic analysis, as is the case with other extant manuscripts similarly marked with stains or finger-prints.[30] Moreover, this tallies with Boyle's apologies about accidents which befell manuscripts in the laboratory. For instance, in a letter of 1 August 1682 to Narcissus Marsh, Provost of Trinity College, Dublin, he apologised about the fate of a letter from Marsh to him which was handed to a servant

[26] See above, pp. 125–6. Note also the movement of vol. 2 of *Experimenta et Observationes Physicae* from one of the boxes on 21 Janury to the green bag by 17 September.

[27] Michael Hunter, *Robert Boyle by Himself and his Friends* (London, 1994), p. 88.

[28] See OED, which records aphetic forms like this prior to the adoption of the current form, of which the earliest instance given dates from 1706–7.

[29] See R.E.W. Maddison, *The Life of the Hon. Robert Boyle* (London, 1969), pp. 128–9, 134, 136.

[30] BP 8, fols. 87v–8 (the content appears to comprise extracts from travel books). Cf., for example, BP 25, pp. 51–4; BP 36, fols. 20–1; MS 188, fols. 61v (a fingerprint), 67–73.

who 'was then busied about the fire, to make a Chymical Experiment I had order'd him to attend in my absence . . . a kindl'd coal unluckily lighted on your Letter, & burnt it quite thorough in that part that contain'd (as I conjectur'd) some of the most important passages of it'.[31] A more elaborate apology of a similar nature appears in the strange broadsheet *Advertisement* which he had printed in May 1688 *about the Loss of many of his Writings: Address'd to Mr* J.W. *to be communicated to those Friends of His, that are* Virtuosi, *which may serve as a kind of Preface to most of his Mutilated and Unfinish'd Writings*.[32] In it he explained how

> One whom I had order'd to do something with a Bottle of Oyl of Vitriol, unluckily broke the Glass just over a flat Chest of Drawers, which I had purposely caus'd to be made to no other use, than to keep in it my own Manuscript Papers, whereof it had then good store. And though I happen'd to be at that time in the Room, and made hast to unlock the distinct Drawers, and take them out; yet the highly corrosive Liquor had made such haste, and such havock, that several Manuscripts, and among them some that I most valu'd, were quite spoil'd, insomuch that there remain'd not words enough undefac'd to declare what Subjects they concern'd; and that the other Manuscripts that mischievous Liquor had reach'd to, had some of them their Leaves half consum'd, and others a greater or lesser part of them; and all that the *Menstruum* touch'd, it made so rotten, that notwithstanding all our diligence, what was once wetted, could never be retriev'd.

The chest in the laboratory is not mentioned in any of the inventories that survive from Boyle's later years. Hence this must have been another of the receptacles for manuscripts other than the core ones which recur in those, and it must have been in these that he stored material other than that linked to treatises on which he was working at the time, including at least some of his voluminous experimental notes. It must also have been in these peripheral locations that he stored material from earlier phases in his intellectual career which still survives yet which does not appear in any of the inventories at all. That this is the case is suggested by the almost random way in which occasional earlier writings appear in the extant inventories, as if Boyle from time to time pulled one out and moved it to one of his main repositories.[33]

If we try to penetrate back to understand Boyle's document management at previous stages in his life, we are handicapped by the fact that, though a handful of lists of Boyle's writings survive, none of these takes the form of an inventory.

31 Royal Society Boyle Letters, 1, fol. 145 (*contain'd* is altered from *concern'd*); printed in Birch (ed.), *Works* (n. 19), vi, 604.

32 For a copy, see BP 36, fol. 51; it will be printed in the 'Pickering Masters' *Works of Boyle*, vol. 11. See also Birch (ed.), *Works* (n. 19), i, ccxxii–iv.

33 See, for instance, the 'Peroration to Pyrof[ilus]' in the list of c. 1680; or the 'Essay of improbable Truths' in the lists of 3 July and 17 Sept. 1691 (this had appeared in 'The Order of My Severall Treatises' in the 1660s: see below); or 'Of the Discernment of Suppositions' in the list in MS 186, fols. 33–4 (see above, n. 171 this had appeared in the lists of Boyle's 'Tracts, &c.' in the 1660s, BP 36, fol. 91; note also MS 189, fols. 38v–9, which refers to Boyle's coming across material relating to this work in a bundle of 'old draughts of Prefaces' which stimulated him to try to reconstruct it).

An example is a famous list of 'Materials & Addenda' for Boyle's treatises dated 25 January 1650, just at the time when he was turning his attention to science.[34] This illustrates the main components of his literary output at that time, dominated by moralistic and religious works, while it also includes a more inchoate category of 'essays', which contain the germ of many of Boyle's later writings, not least on scientific topics. From these beginnings developed the burst of activity as a scientific writer which occurred during the years that Boyle lived at Oxford from 1655 to 1668 and which established his reputation throughout Europe.

Our understanding of the appearance and organisation of Boyle's papers at this stage in his career is further inhibited by the extent to which material was subsequently discarded, particularly the manuscripts of books that Boyle published from 1659 onwards, a decision that is probably attributable to Boyle himself, who presumably saw such material as superfluous once it had got into print. Thus, for instance, we have virtually no manuscript material at all related to Boyle's *New Experiments Physico-Mechanical, Touching the Spring of the Air and its Effects* of 1660, or his *The Origin of Forms and Qualities* of 1666.

On the other hand, manuscript material *does* survive relating to certain books of this period, and the same is also true later in Boyle's career. From the former, worthwhile conclusions can be drawn about the form that Boyle's writings took during the formative stages of his career. A number of Boyle's moralistic writings from his adolescence survive since they were never published at all, while he also kept sections of treatises on natural philosophy from his Oxford period which, though published, were not published in full: a case in point is his *Some Considerations touching the Usefulness of Natural Philosophy*, of which parts were published in 1663 and 1671, but of which significant parts never saw the light of day. (Later, a similar state of affairs is in evidence with Boyle's *Free Enquiry into the Vulgarly Receiv'd Notion of Nature* (1686): here, too, drafts for published sections were kept interspersed with unpublished material, evidently because it was too much trouble to sort the two out.)[35]

The manuscript material relating to *The Usefulness of Natural Philosophy* is particularly significant, since it reveals a process of evolution which itself provides the background to Boyle's disposition of his papers in his later years.[36] All of the early works by Boyle that survive – from his adolescent exercises in the late 1640s to the writings of his early years at Oxford in the mid-1650s – show a common format. They were written in continuously paginated paper-books, most often of foolscap format, with a pagination often running to over 100 pages. Hence, at this

[34] BP 36, fol. 86, printed in Maddison, *Life* (n. 29), p. 64, and in John T.Harwood (ed.), *The Early Essays and Ethics of Robert Boyle* (Carbondale and Edwardsville, 1991), p. xix. For the context, see Michael Hunter, 'How Boyle Became a Scientist', *History of Science*, 33 (1995), 59–103, esp. pp. 67–8.

[35] The survival of manuscript material relating to Boyle's published works will be tabulated in full in the forthcoming 'Pickering Masters' edition of *The Works of Robert Boyle* by Michael Hunter and Edward B. Davis. That for *Notion of Nature* is discussed and tabulated in Hunter and Davis, 'Making of Boyle's *Free Enquiry*' (n. 9).

[36] For documentation of the points made in this and the following paragraphs, see ibid., pp. 214f.

point, his archive could quite easily predominantly have taken the form of a series of discrete volumes on a shelf.

However, in the late 1650s – during the period when he was working on *Usefulness* – Boyle's method of composition changed. Now, he started to write his treatises in separate, relatively brief sections on folded sheets of paper, interleaving them or stitching them together to give him the number of pages that he required. The advantage of this was of flexibility: it is evident from the surviving sections of the manuscript of the original version of *Usefulness* that he found the old format of a lengthy, consecutively paginated text increasingly constricting when he wished to alter and extend a work by adding extra examples or digressing on relevant points. The new format, on the other hand, meant that the structure and content of a work was almost infinitely adaptible, since sections could be extended, recombined and intercalated almost at will. If Boyle's writings may up to this time have comprised discrete volumes, now they must increasingly have comprised heaps of disparate, separately composed sections. What is more extraordinary still is how few of Boyle's manuscripts from the mid-1650s onwards have any original pagination at all, being linked to the sheet following them by a catchword and nothing else. Hence one may speculate that it was *this* that necessitated his keeping the papers which made up his different books in discrete folders tied up with strings, a practice which probably originated in connection with this change in his method of composition.

Indeed, although composition in separate sections gave Boyle flexibility, it also brought with it attendant dangers of disorganisation which had not occurred with continuously paginated texts. It is not surprising that problems of document management arose, to which Boyle testifies in various comments in the prefatory material to his later writings, perhaps most strikingly in the 'Advertisement' prefixed to his *Free Enquiry into the Vulgarly Receiv'd Notion of Nature* (1686), where he had to apologise for the fact that certain parts of the work were actually printed in the wrong order due to accidental 'Transpositions of the loose Sheets, wherein (and not in a Book,) the Copy was sent to the Press'.[37] It is also clear that material went astray, with serendipidous side effects: Boyle tells on one occasion how, 'Rumaging the other day among my old writings in search of one I could not finde I lighted on another I did not seek for.'[38]

Moreover, though certain sections of text clearly belonged to a specific work and could be classified accordingly, as his life went on Boyle was increasingly left with a residue of incomplete fragments. Indeed, he gives some sense of the variety of materials that existed in an unpublished prefatory passage which it is worth quoting here at length.[39] Apologising for the state of his imperfect writings, he explained how

[37] Ibid., p. 227. The published version will be printed in the 'Pickering Masters' *Works of Boyle*, vol. 10.
[38] MS 187, fol. 148.
[39] MS 187, fols. 129v–30. The last nineteen words are written in the margin.

> I thought fit to <cast>[40] them into 4 or 5 difering sorts. The 1st was the consist of greater fragments or <un>entire pieces <each>[41] of them containing several sheets or at lest Pages about the subject it treats of. Another sort consisteth of lesser Fragments being of the same nature or Condition in [sic] the former but inferior to them in bulk. The 3d consisted of imperfect or unfinisht Pieces in which thô the beginning & perhaps som other part (one or more) of each[42] distinct Paper as [sic] uniform enough connected & perhaps pretty large, yet the coherence were somtimes interrupted & to be sure the Discourse was <incompleat or> or unfinisht. A 4th sort was a Chaos of Papers relating to som grand or comprehensive subject, as Divinity, Philosophy, Physick, Chymistry, &c. And the 5th was a Mere Chaos or Pandets receiving Fragments for the most part a smaler kind of all sorts of subjects thô here & there som that chancd to be numerous enough & somwhat near of kin were pyld up together or cast into distinct Heaps of Rubish.

References to this 'Chaos', which clearly comprised one of the peripheral components of the archive beyond the thematic folders at its core, recur in various of the inventories. Sometimes, these distinguish between 'greater' and 'lesser' fragments (though how successfully is unclear). In addition, they sometimes differentiate between scientific and theological material, and it is here worth drawing attention to the fact that Boyle's entire classificatory system was predicated on a distinction between papers dealing with 'theology' and 'philosophy', to which he sometimes added subsidiary divisions relating to chemistry, medicine, and miscellaneous materials, as in a memorandum evidently of things to do prior to his death, which includes the entry: 'oversight of my writings, Theol[ogical], Philos[ophical,] Chymicall, Medic[al] &c'.[43]

A further development in Boyle's later years was the use of notebooks as a repository for drafts, especially of shorter passages intended for the various treatises that he was working on at any particular time; he then had these copied out onto sheets of paper to join the texts to which they belonged.[44] These notebooks could evidently be acquired ready-made, complete with strings to tie them up, another example of the stationery trade already referred to. As with folders and ribbons, a choice of colours was available, which helped to identify different items. For instance, there is a reference in one such book, now rebound and long without any ties, to 'the following Note booke with green Strings'.[45] Puzzingly, none of these notebooks appear in the inventories, but, since they were apparently bound volumes, they may not have been stored with the folders and bundles of papers in

40 Replacing *sort* deleted. The next word but one, *into*, is altered from *to*.

41 Replacing *of a considerable bulk* deleted.

42 Followed by *the* deleted. In the next sentence, *Chaos* was altered in the course of composition.

43 See BP 36, fol. 178. See also MS 189, fol. 67v (a key to abbreviations, copied in MS 186, fol. 1v), and BP 36, fol. 87 (a memorandum similar to fol. 178): the latter is cited in Harwood, *Essays and Ethics* (n. 34), p. xviii n. 7, though he fails to note that the five categories he itemises are preceded by a sixth, 'The Chachamick Papers' [sic: this concept does not appear in OED and has defeated me].

44 Hunter and Davis, 'Making of Boyle's *Free Enquiry*' (n. 9), pp. 218–19.

45 MS 189, fol. 67v (copied in MS 186, fol. 1v).

the boxes in Boyle's rooms, but may have been shelved with his books (Evelyn states that Boyle's library was in a discrete room).[46]

So far, I have principally been talking about the component of the archive that relates to Boyle's activity as a writer of the treatises on scientific and religious topics by which he established his fame. However, I want to end by talking about another, equally significant component that existed alongside this, and this is a series of banks of data, work-diaries if you like. These date back to Boyle's earliest years: indeed, the earliest are literary in content, though from 1650 onwards they change abruptly to a record of recipes and pieces of information that he was given by others, juxtaposed with experimental findings and observational data of Boyle's own.[47] These provide a sequential account of the raw data that Boyle acquired throughout his life, and clearly he valued them both for their record of the profuse and inventive experimental trials to which he devoted so much of his time, and because they contained – in the words of the 1684 inventory – 'things that I learn'd in conversing with Pilates, sea-captains, and other Persons that have travail'd unto the Indies, or other remote country in Order to make Physica peregrinare'. Boyle thought of these as comprising units of 100, and each time he had recorded a 'century' of such observations he began a new sequence. By the end of his life he was onto his eighteenth century (the latest dated experiments are from August 1691[48]), and substantial quantities of these records survive. On the other hand, there were evidently once more, since Boyle's 1688 broadsheet was partly inspired by the loss of 'four of five Centuries of Experiments of my Own, and other Matters of Fact, which from time to time I had committed to Paper, as they were made and observ'd', together with 'seven or eight Centuries of Notions, Remarks, Explications and Illustrations of divers things in Philosophy, which I had committed to Writing as they chanc'd to occur to my Thoughts', which in a related manuscript he noted he 'strongly suspected to have been surreptitiously conveied away by some' who 'expected to find valuabler things, than I supposd they met with'.[49]

Clearly the intention was that this data, although sequentially recorded by date, was to be deployed more thematically in connection with Boyle's writings on scientific subjects. Indeed, a relatively early example of these collections, dating from the early 1660s, is specifically entitled 'Philosophicall Entrys & Memorialls (of all sorts,) Here confusedly throwne together; to be Hence transferr'd to the Severall Treatises whereto they belong'.[50] Moreover, up to a point, it is possible to document Boyle making piecemeal use of this data in his writings, for instance in his *Experiments and Considerations about the Porosity of Bodies* (1684), which more than once includes material of this kind, sometimes even transcribing the

46 Hunter, *Boyle by Himself and his Friends* (n. 27), p. 88.
47 Hunter, 'How Boyle Became a Scientist' (n. 34), pp. 66–7. For later examples, see Hunter, *Letters and Papers* (n. 1), passim. An electronic edition of this material is currently being prepared.
48 BP 21, pp. 219–54 (p. 254).
49 BP 36, fol. 6, printed in Birch (ed.), *Works* (n. 19), i, cxxv–vi. *Surreptitiously* was originally written incorrectly and has been altered by Boyle.
50 BP 22, p. 1.

date of the experiment.[51] There is arguably a parallel here with the commonplace books of figures like Evelyn, though appropriately unbookish for a figure like Boyle whom Evelyn himself described 'as learning more from Men, Real Experiments, & his Laboratory (which was ample and well-furnish'd), than from Books'.[52]

On the other hand, it is clear that Boyle hoped to sort and redeploy this material more systematically, and various strategies are in evidence, starting in the 1660s. At that point, Boyle produced a whole series of copies of a document called 'The order of my severall Treatises'.[53] In this, each of his published or unpublished works was given a letter and a number, as if with the intention of going through his notes marking each item with a reference of this kind, and then having the information redistributed by having it recopied in separate series. In addition, in two copies a further series of numbers were given, which could conceivably refer to a series of files, since in some cases more than one number is attached to a single work. There is also a verse mnemonic, aimed to fix in the memory just which letter stood for which work: for instance

E calls the *Printed usefullnesse* to mind
Tryalls of Colours are in *F* combin'd.[54]

Later, Boyle devised a variant on this theme, entitled 'Paralipomena'. Again, this comprised a list of letters and numbers denoting headings under which material was to be classified and again multiple copies survive among the Boyle Papers, while by this stage an intention to make a book out of this material is implied by the description of the categories as 'Chapters' and the provision of prefatory material.[55]

What is puzzling, however, is that it is very difficult to find evidence of Boyle putting the strategies indicated in these much recopied documents into practice. Only a handful of items in his profuse data collections have endorsements in the form of numbers or letters which can be correlated with these lists, and I have not come across any collections of redistributed data.[56] Instead, Boyle seems to have tried other methods, such as writing the title of the treatise to which an item was relevant in crayon in the margin beside it, or summarising its content, or compiling numbered tables of his different 'centuries' of experiments, and then filling them

51 See the forthcoming 'Pickering Masters' *Works of Boyle*, vol. 10.

52 Hunter, *Boyle by Himself and his Friends* (n. 27), p. 88.

53 One of these, from BP 8, fol. 64v, is printed in Rose-Mary Sargent, *The Diffident Naturalist: Robert Boyle and the Philosophy of Experiment* (Chicago, 1995), p. 140, who refers to a second on pp. 282–3 n. 46. However, there are in fact a number of slightly different recensions scattered through the archive; these will be fully tabulated in the 'Pickering Masters' *Works of Boyle*, vol. 14.

54 BP 27, fols. 2v–3.

55 See esp. BP 25, pp. 217f. See also Sargent, *Diffident Naturalist* (n. 53), pp. 140–1 and 283 nn. 47–51. Again, different versions survive, which I intend to analyse in a future study linked to my electronic edition of the work-diaries.

56 There are, however, some classified listings which I will deal with in the forthcoming study referred to in the previous note.

in with notes as to which were relevant to which work, or conducting a similar exercise in relation to a specific treatise.[57] Only right at the end of his life did Boyle publish a brief volume comprising data derived from these sources, roughly categorised by topic, his *Experimenta et Observationes Physicae* of 1691, to which his prefatory remarks to his 'Paralipomena' imply that that was to be seen as an appendix.[58]

Otherwise, however, the multiple copies of the elaborate classificatory system outlined in 'The order of my severall Treatises' and the 'Paralipomena' from the 1660s onwards represent an unrealised ambition to give an ordering to a component of the archive that otherwise remained less useful than it might have done due to its large scale and heterogeneous content. Clearly, there was a failure of document management here, and I make no apology for drawing attention to it, since – as in similar cases which have recently been studied, for instance Locke's commonplace books, or Evelyn's library catalogue[59] – there is as much to be learnt from such failures as from successes.

In Boyle's case, this one was arguably syptomatic of an unresolved tension within his natural philosophy as a whole. Though Steven Shapin and Simon Schaffer have claimed in *Leviathan and the Air-Pump* that 'matters of fact' were 'the foundational category' of Boyle's scientific practice,[60] this is a truism which elides a problem with which such data presented him. For all the value that he attached to the data that he collected with such Baconian enthusiasm, Boyle was convinced that it was only useable if given a shape.[61] Indeed, he was embarrassed by the shapelessness of such miscellaneous collections: the attempted ordering by theme in his 'Order of my severall Treatises' and 'Paralipomena' has a faint echo in the published *Experimenta et Observationes Physicae*, which has experiments roughly grouped by subject and published in multiples of five, many of them entitled 'Pentades'. Yet Boyle's difficulties are apparent even from the delayed and incomplete nature of that publication – it had been projected since the 1670s, as witness its introductory letter to Oldenburg – and from his failure to mould his massive arsenals of data into the promised sequels to it.

In this paper I have concentrated on the core of the Boyle archive, ignoring such items as Latin translations of his works, copies of works by others, and suchlike

57 For examples of each of these practices, see respectively BP 17, fols. 147f., BP 26, fols. 78f. (titles of works in crayon or pencil); BP 8, fols. 63f., BP 21, pp. 191f. (brief notes on content); BP 27, pp. 197f. (numbered list, classified according to work); BP 10, fols. 113–14 (titles of works, with list of numbers, presumably of relevant experiments, annexed).

58 BP 25, pp. 217f.; Birch (ed.), *Works* (n. 19), v, 564f. The history of this work, including Boyle's plans for it from the 1670s onwards and his intention to publish a sequel which failed to materialise, will be fully documented in the forthcoming 'Pickering Masters' *Works of Boyle*, vol. 11.

59 See above, pp. 13–14.

60 Steven Shapin and Simon Schaffer, *Leviathan and the Air-Pump: Hobbes, Boyle and the Experimental Life* (Princeton, 1985), p. 73.

61 See Michael Hunter, 'The Reluctant Philanthropist: Robert Boyle and the "Communication of Secrets and Receits in Physick" ', in Ole Peter Grell and Andrew Cunningham (eds.), *Religio Medici: Medicine and Religion in Seventeenth-century England* (Aldershot, 1996), pp. 247–72, esp. p. 259.

penumbra, which only intensified the impression of disorder for those who inherited his manuscripts. But what I hope should be clear is that the central components of the archive which I have described here had both positive and negative characteristics. The boxes containing colour-coded folders for Boyle's work in progress represented a highly flexible method of archival storage, even if it was dependent on an intervening principle in the form of Boyle's own mind, and easily seemed chaotic when this was removed. But Boyle's profuse data-banks represented more of a problem, as is testified by the abortive schemes for organising them which the archive also contains. Perhaps it is not fanciful to see the archive itself as a mirror of Boyle's strengths and weaknesses as a thinker.

9

A 'connected system'? The Snare of a Beautiful Hand and the Unity of Newton's Archive

Rob Iliffe

> ... however imperfect some of these papers may be, yet certainly they must contain something valuable to the publick, being the works of Sir Isaac Newton.[1]

Amongst the archives of early modern scholars which have been preserved, it is arguable that the intellectual legacy of Isaac Newton has received more attention than that of anyone else. However, few scholars were given the opportunity to fully examine the Portsmouth Papers (as the archive was then known) until 1872, when the 5th Earl of Portsmouth, prompted by the Duke of Devonshire, generously gave permission for the University of Cambridge to identify and then take away the 'scientific' portion of the papers. When the Committee set up to examine and classify the papers drew up their report sixteen years later, the 'non-scientific' and apparently uninteresting items were returned to Hurstbourne Park in Hampshire, where they survived a serious fire in 1891. Thereafter scholars had easier access to Newton's scientific work, although manuscripts indicative of his other intellectual interests (while broadly known) were not generally available for examination until they were sold off in the 1930s, splitting the archive into even smaller portions. The increasing availability of these papers for analysis has been accompanied by a number of projects aimed at producing editions of Newton's scientific work. Following a renewed interest in Newton after the Second World War, scholars now have access to an authoritative edition of his correspondence and a comprehensive selection of his mathematical papers, while the first of a projected three volume edition of his optical papers and letters appeared over a decade ago. For a fair sum, individuals with access to a microfilm reader can now take advantage of the Chadwyck-Healey edition of Newton's papers in order to have at their disposal virtually the entire extant Newton manuscript collection.[2]

I would like to thank Larry Stewart and Michael Hunter for comments on previous versions of this paper.

1 John Conduitt, remarks on manuscript draft entitled 'Original of the Kingdome of Egypt. Chap II', May 1729; New College, Oxford, MS I, fols. 88–90.

2 Amongst the most important editions or publications of portions of Newton's archive are H.W. Turnbull, J.P. Scott, A.R. Hall and L. Tilling (eds.), *The Correspondence of Isaac Newton* (7 vols., Cambridge,

Inevitably, modern views of Newton have been coloured if not determined by the attitudes of those who have commented on his work since his death. In recent times, for example, as more evidence came to light in the 1960s and 1970s of his religious, alchemical and philosophical interests, a brand new Newton emerged, in many ways utterly different from the rationalist physicist and mathematician celebrated in the two centuries after his demise.[3] Two issues arise from this re-evaluation of Newton's intellectual interests. If a radical distinction is posited between his 'scientific' and 'non-scientific' works, there is a question of how to relate these areas. One can either concentrate on one part of Newton's work to the exclusion of others or one can attempt to view parts of his work in terms of his broader efforts. While the latter seems to have certain advantages, it has invariably borne an attendant historiographical obligation to show how for Newton one area was more 'important' than or 'influenced' another. This is further complicated by the fact that our own academic disciplinary categories, even when they appear similar, do not neatly correspond to those of the earlier period. This being the case, then finding connections or distinctions between different regions of his work based on our own taxonomy of disciplines can be an irrelevant exercise, although the dichotomy between 'scientific' and 'non-scientific' pursuits is difficult to avoid. In her second book on Newton's alchemy, for example, Betty Dobbs argues for a deep connection between the alchemical notion of attraction and the concept of Universal Gravitation, although her depiction of Newton's work as 'Janus-faced' re-inserts the *a priori* distinction between alchemy and mechanics which she criticises elsewhere. Dobbs solves the problem by asserting the fundamental coherence of the author of the manuscripts attributed to Newton: 'The Janus-like faces of Newton were after all the production of a single mind, and their very bifurcation may be more of a modern optical illusion than an actuality.' Whatever our prejudices, Newton himself was 'convi[nced] of the unity of Truth'.[4]

The opportunity granted by microfilm of bringing together nearly all the

1959–77); D.T. Whiteside (ed.), *The Mathematical Papers of Isaac Newton* (8 vols., Cambridge, 1967–81); A.R. Hall and M.B. Hall (eds.), *Unpublished Scientific Papers of Isaac Newton* (Cambridge, 1962); J. Herivel, *The Background to Newton's 'Principia'. A Study of Newton's Dynamical Researches in the Years 1664–84* (Cambridge, 1964); A.E. Shapiro (ed.), *The Optical Papers of Isaac Newton* (3 vols., Cambridge, 1984–); B.J.T. Dobbs, *The Janus Faces of Genius* (Cambridge, 1991), pp. 256–305. See also below, n. 28. I.B. Cohen, *Introduction to Newton's 'Principia'* (Cambridge, 1971) is a detailed analysis of the manuscripts related to Newton's greatest work while H.J. MacLachlan's flawed *Sir Isaac Newton's Theological Manuscripts* (Liverpool, 1950) remains the only extended edition of Newton's theological papers.

3 Significant works for reevaluating the relatively neglected portions of Newton's intellectual endeavour were J.E. McGuire and P.M. Rattansi, 'Newton and the Pipes of Pan', *Notes and Records of the Royal Society*, 21 (1966), 108–43; D.C. Kubrin, 'Newton and the cyclical cosmos: providence and the mechanical philosophy', *Journal of the History of Ideas*, 28 (1967), 325–46; J.E. McGuire, 'Force, active principles and Newton's invisible realm', *Ambix*, 15 (1968), 154–208; F.E. Manuel, *A Portrait of Isaac Newton* (Cambridge, Mass., 1968); id., *The Religion of Isaac Newton* (Oxford, 1974); B.J.T. Dobbs, *The Foundations of Newton's Alchemy, or, 'The Hunting of the Greene Lyon'* (Cambridge, 1975); K. Figala, 'Newton as alchemist', *History of Science*, 15 (1977), 102–37; and D. Castillejo, *The Expanding Force in Newton's Cosmos* (Madrid, 1981).

4 See Dobbs, *Janus Faces* (n. 2), pp. 3–5, 6, and I.B. Cohen, 'The *Principia*, Universal Gravitation, and the "Newtonian style" in relation to the Newtonian Revolution in science: notes on the occasion of the 250th anniversary of Newton's death,' in Z. Bechler (ed.), *Contemporary Newtonian Research* (Dor-

available Newtonian archival material seems to offer one way of avoiding interpretations which are prey to the tides of fashion. No longer are items suppressed because of the moral or theological nature of their content, and in terms of ideology, the microfilm collection comes as close as possible to being guileless. Yet anyone who wishes to offer a broad account of Newton's intellectual endeavours is immediately confronted by the sheer mass of manuscript evidence, and the archive is still divided into sections which reflect the interests of those who have classified it in the major archival upheavals since 1727. It is important to be aware of this and to note that these interests have not always privileged the 'scientific' portions of his work. Indeed, reflecting the concerns of a different age, early interest in Newton's legacy was largely directed to what the papers might say about the theological beliefs which he had so skilfully concealed in his lifetime. In the rest of this paper I examine the fate of Newton's manuscripts and suggest that a reconsideration of the history of this archive throws up some questions relating both to his working habits and to his intellectual approach as a whole. How, for example, did Newton himself divide up his intellectual world, and what can be gained from asking whether there were significant connections – or differences – between these spheres? Given his professed reluctance to publish anything lest disputes should arise, what can be said about his possible motivations for writing anything at all? What did he believe was the real or putative audience for his works, and to what extent were his more complete texts circulated amongst right minded 'friends'? If publication was not his purpose in writing, what was the nature of the fascinating evolution of various texts or 'projects' over the period from the 1660s to the 1720s? What, finally, do these considerations tell us about who Newton thought he was?

Newton was engaged on a number of different projects, which were carried on simultaneously over many decades and many of which were connected to others; more generally, I argue that he divided up his intellectual world into two spheres. However, despite evidence of intellectual fragmentation, the consideration of Newton's entire intellectual life through his archive can point to a larger picture in terms of which individual projects can be grasped in a more nuanced manner. As Thomas Söderqvist has suggested, such an integrated view of an individual should pay more attention to the 'non-dramatic', quotidien regularities of a life rather than to the 'outward contributions' which usually constitute history. This can be done by looking both at the sources upon which Newton drew for individual projects, as well as at the kinds of questions and audiences he was attempting to address in each case. Looking for similarities across a broad range of intellectual interests is hardly the same as positing some transcendental 'essential' subject underlying Newton's work, while pointing to differences, especially when they can be shown to be those endorsed by the author of the manuscripts in question, is not to deconstruct the subject beyond repair.[5]

drecht, 1982), 21–108. For a critique of ways in which historians have used interpretive strategies as fragmentation, coherence and consistency in order to link or disaggregate portions of Newton's *oeuvre*, see S. Schaffer and S. Shapin, 'Making Newton: on interpreting scientific texts' (unpublished typescript, 1985).

5 For an existential notion of 'project' see Thomas Söderqvist, 'Existential projects and existential choice

The fate of the archive

Newton died intestate on 20 March 1727 and there followed a series of family squabbles over how best to dispose of his manuscripts for financial benefit. John Conduitt recorded that there were 'several contests & disputes in the Prerogative court about the Administration & the disposal of the manuscripts . . .' and, acting on behalf of his wife Catherine (daughter of Newton's step-sister Hannah), Conduitt was the only person able to post a bond for Newton's debts, on which grounds he claimed entitlement to all of the unpublished material. According to his memorandum, as part of the agreement with Newton's administrators it was determined that the different factions vying for a share of the archive should be allowed to peruse the 'papers and Manuscripts' and 'such as are treatises be afterwards examined by Dr Pellett & printed if thought proper by him & the Adm[inistrato]rs & sold to the best advantage, & in the meantime be loged with Mrs Conduitt'. Conduitt was also allowed to guard the papers not thought fit to be printed, giving bond of £2000 'that if he ever he shall publish any or make any advantage thereof' he was to be accountable for them. The relatives inspected all the papers and manuscripts at Newton's house and 'took Catalogues of them', and having examined them, the appraisers Comyns and Ward 'thought only the Chronology and Prophecies fitt to be appraised & sett the value of £250 upon the Chronology, & no value upon the Prophecies they being imperfect'. The inventory also noted the existence of 'Manuscripts in a box sealed up at the house of John Conduitt Esq.' as well as 'several boxes at Sir Isaac Newton's house wherein are contained many loose papers and Letters, relating to the office of the Mint, the manuscripts above mentioned and his Mathematical Works already published and likewise two small parcels of papers of the same kind in the box att the house of John Conduitt Esq.'[6]

Thomas Pellet drew up an inventory on 20, 22 and 26 May 1727 and listed 81 separate items of which 37, or nearly half, were concerned with chronology or theology. Pellet decided that only five manuscripts were 'fit to be printed', viz. No. 80, 'Containing an abstract of the Chronology being 12 half sheets in folio & the Chronology being 92 half sheets in folio'; No. 33, 'de Motu Corporum' or *liber secundus*, in 56 half sheets in folio; No. 81, '40 half sheets in folio being the History

in science: scientific biography as an edifying genre', in M. Shortland and R. Yeo (eds.), *Telling Lives in Science: Essays on Scientific Biography* (Cambridge, 1996), pp. 45–84, esp. pp. 58 and 74. For the 'myth of personal coherence' see J. Clifford, ' "Hanging up looking glasses at odd corners": ethnographical prospects', in D. Aaron (ed.), *Studies in Biography* (Cambridge, Mass., 1978), pp. 41–56.

6 Keynes MS 127A and Whiteside, *Mathematical Papers*, i, xvii–xx; cf. ibid., p. xix n. 9, for the inventory made by Comyns and Ward on 18th April. Newton's library was valued at £270, and the Warden of Fleet Prison, John Huggins, paid £5–5s as a deposit for it on 2 June 1727, 'Save only and except that the books particularly mentioned in the Page [i.e. of the same document] to have Sir Isaac Newton's notes, are not to be delivered to the said Mr Huggins, but to be valued by two indifferent persons . . . and such valuation to be deducted out of the Three hundred pounds'; see J. Harrison, *The Library of Isaac Newton* (Cambridge, 1978), p. 31 n. 1. In addition to the deposit, Huggins paid £300 for the library on 20 July, and the six books with Newton's annotations were returned to the Conduitts; ibid., pp. 33–4, esp. p. 33 n. 2.

of the Prophecies in 10 chapters & part of the 11th unfinished'; No. 38, '31 half sheets in folio being paradoxical questions concerning Athanasius', and No. 61, 'An imperfect mathematical tract'. Although Pellet's language makes it look as if he censored papers according to an ethical whim, his brief was to uncover works that might be considered complete and hence ready for publication (the rest being 'not fit to be printed'), and only two of these texts were deemed in a requisite state. Indeed, the last three were initially only to be 'reconsidered' for printing, but only No. 81 saw the light of day in the eighteenth century, with the heterodox nature of Newton's work on Athanasius accounting for the non-appearance of No. 38. However, as we shall see, a number of people were increasingly interested in printing unpublished Newton material.[7]

Evidence from this list for the original order of Newton's papers is extremely sketchy, and beyond fitting manuscripts together and reordering papers from internal evidence there is almost no way of finding out how Newton classified his papers (one can of course argue to Newton's concerns from the type of papers which do survive). On one of the manuscript drafts by Conduitt there is a reference to cataloguing having reached number 72 by 13 May 1727, and Nos. 73 to 81 are described as being 'In 7 boxes & a hart trunk', suggesting that each item was boxed after or even before inspection. With regard to these, the nature of theological and chronological items is now more readily identifiable than the mathematical or philosophical portions, since for most items one can easily collate the title and the number of folios in each lot with the list compiled for Sotheby's sale of the Portsmouth Papers in 1936. The next evidence comes from the short catalogue compiled by Samuel Horsley and William Mann Godschall after a visit to Hurstbourne in October 1777, although in the meantime, following a codicil made by Catherine Conduitt in 1737, a batch of theological papers had been despatched by Lord Portsmouth to the pro-Newtonian divine Arthur Ashley Sykes in November 1755.[8]

The papers had come into the possession of the Portsmouth family following the marriage of Catherine Conduitt (John Conduitt's daughter) to John Wallop

7 Whiteside, *Mathematical Papers*, i, xix–xxii; Pellet examined nos. 1–20, 21–40 and 41–81 on the respective dates. This list is extant in two drafts in Conduitt's hand; Keynes MS 127A (4). No. 80 was sold to Jacob Tonson for £350, published as *The Chronology of Ancient Kingdoms Amended* (London, 1728) and is now CUL Add. MSS 3988 and 3987; No. 33 was sold to Tonson for £31 10*s* and published as *De Mundi Systemate Liber Isaaci Newtoni* (London, 1728), and is now CUL Add. MS 3990; No. 81 was printed as *Observations upon the Prophecies of Daniel, and the Apocalypse of St. John. In Two Parts*, (London, 1733), and is now CUL Add. MS 3989; No. 38 remains unpublished as Keynes MS 10 and No. 61 is probably the combined 1666 English and 1671 Latin fluxional tracts (now CUL Add. MSS 3958 fols 48v–63v and 3960.14); see Whiteside, *Mathematical Papers*, i, xx n. 15.

8 Keynes MS 127A(4), 'An Acc[ount] of Sir Isaac Newton's papers – sent to the Rev. Dr. Sikes in London', consisted of 'No (1) A bundle of Papers on the Revelation & Daniel compleat', 'No 2 A bundle of papers – Complete Chapters of the Host of Heaven &c 6 papers folded which are Duplicates Relating where they are put', 'No 3 A bundle of papers of Loose Independent pieces of little use – unless any may afford any hint in writing Sir I.'s life', 'No 4 A bundle of Papers on the Revelation some part duplicate', 'No 5 A bundle of papers of no Connection or use being only Memorandums which Sir Isaac made out of Book's', 'No 6 papers of Jewish Synagogues & Christian Churches (loose papers)', 'No 7 Proemium to Church History &c a bundle', 'No 8 A book of Chronology &c', 'No 9 Duplicates & papers relating to the time – Imperfect', 'No 10 An Abstract, or first thoughts on the Host of Heaven' and 'No 11 on the Revelation, some little duplicates'. These are now Bodleian Library, New College MS 361.1–4.

(Viscount Lymington after 1743) in 1740, and in 1755 Sykes had compiled a 'digest' of Newton's religious and chronological papers according to the terms of Catherine Conduitt senior's codicil. This asserted that the executor should 'lay all the Tracts relating to Divinity before Dr Sykes . . . in hopes he will prepare them for the press . . . all of them I ordain shall be printed and published, so they be done with care and exactness'. Sykes died before he could fulfil his task and the papers were passed on to the Reverend Jeffrey Ekins, in whose family the papers remained until they were given to New College, Oxford in 1872.[9]

Alongside Catherine Conduitt, a number of contemporaries expressed the hope that Newton's theological papers would be published although the unorthodox nature of his religious views had already been widely discussed. For example, John Craig told John Conduitt soon after Newton died that the latter had expended great time and effort on theology and ecclesiastical history

> & had drawn up in writing great collections out of both; &, to show how earnest he was in Religion, he had written a long explication of remarkable parts of the old & new Testaments, while his understanding was in its greatest perfection, lest the infidells might pretend that his applying himself to the study of Religion was the effect of Dotage . . . But now its hoped that the worthy and ingenious Mr Conduit will take care that they be published that the world may see that Sir I: Newton was as good a Christian as he was a Mathematician and Philosopher.[10]

Others also heard of the potential significance of Newton's papers, and Colin MacLaurin told the Scottish clergyman Robert Wodrow in November 1727 that after some enquiries to Newton's trustees he had been told that there was 'nothing of any importance found among his papers, save his large Chronology, which is now printing, and will be published . . . there are very great expectations from it'. Wodrow was reassured that there was nothing in the archive that was relevant to the Trinitarian controversy, and despite noting that there had been rumours that Newton had been part of an anti-Trinitarian grouping, he remarked that it was good that 'Sir Isaack hath medled very litle with Divinity throu his life . . .' However, two years later MacLaurin told him that Newton's papers contained some 'peculiar thoughts' on prophecy, and that Newton, whom he had heard 'express himself pretty strongly upon the subordination of the Son to the Father', might have been 'pretty much in Doctor Clerk's sentiments'. With this in mind, it is not surprising that little theological material was published in the two centuries after his death although text which was largely in note form or whose immediate theological purpose had become opaque by the middle of the eighteenth century was also

9 The portion of Catherine Conduitt's codicil is from David Brewster, *Memoirs of the Life, Writings and Discoveries of Sir Isaac Newton* (2 vols., Edinburgh, 1855), ii, 341 n. 3. Sykes defended Newton's *Chronology* against the attacks of critics (in particular William Warburton) *in An Examination of Mr. Warburton's Account of the Conduct of the Antient Legislators, of the Double Doctrine of the Old Philosophers, of the Theocracy of the Jews, and of Sir Isaac Newton's Chronology* (London, 1744), and *A Defence of the Examination of Mr. Warburton's Account* (London, 1746).

10 Craig to Conduitt, 7 April 1727; Keynes MS 132.

rejected. Ekins, for example, claimed in a letter of 27 March 1757 that of the tracts sent to Sykes by Portsmouth, 'None of them are perfect excepte one little tract relating to the controverted text 1. John 5, 7 &c which surreptitiously got into print about two or three years ago. The rest seem very inconsiderable except . . . that the manuscript of Sir Isaac's Chronology is amongst these papers & is pretty fairly wrote.'[11]

Interest was by no means centred only on Newton's theological views, and a small number of mathematical manuscripts and letters on various topics were circulated or published after Newton died. William Jones, who as a young man had brought out a collection of some of Newton's short mathematical works in 1711, had access to Newton's own papers and more significantly to transcripts made over a number of years by John Collins, and he prepared a short history of Newton's mathematical researches when his acquaintance Thomas Birch approached him in the 1730s in connection with research for an article on Newton in an improved version of Bayle's *Dictionary*. Some of Jones's papers were passed around mathematicians in London and further copies were made of the transcripts in his possession. These constituted the best known source for Newton's private mathematical researches, such as those published by Rigaud in 1838, until the Portsmouth Collection was made available in the 1880s.[12]

The greatest interest in Newton's work on natural philosophy came in connection with the efforts to understand both the 'Queries' added to the three major editions of his book on optics and Newton's manuscript 'De natura acidorum', written in 1692 and published in 1710. A number of previously unpublished items from Newton's correspondence were published for the first time in the new edition of Bayle's *Dictionary* in the late 1730s, including the letters between Halley and Newton just before the publication of the *Principia*, together with Newton's correspondence with Flamsteed concerning issues surrounding the comet of 1680–81.[13] Knowledge of Newton's 'private' views on matter theory and the nature of gravity increased dramatically in the middle of the century with the

11 See L. Stewart, 'Seeing through the Scholium: Religion and reading Newton in the eighteenth century', *History of Science*, 34 (1996), 123–65; R. Wodrow, *Analecta: or, Materials for a History of Remarkable Providences; Mostly Relating to Scotch Ministers and Christians* (4 vols., Edinburgh, 1842–3), iii, 461–2 and iv, 59; Ekins to Joseph Wilcox, Trinity College MS R.16.38, fols. 416–17, cited in Whiteside, *Mathematical Papers*, i, xxiv n. 27. The analysis of 1 John 5:7 was the first of two letters to John Locke (now New College MS 361.4 fols. 2–41), the second of which concerned 1 Tim. 3:16. The text to which Ekins referred was *Two Letters of Sir Isaac Newton to Mr Le Clerc* (London, 1754), printed in *Newton Correspondence*, iii, 83–129. Sykes and Ekins were executors of the wills of Catherine Conduitt senior and junior respectively; see Whiteside, *Mathematical Papers*, i, xxiv n. 27. I am indebted to Stephen Snobelen and Paul Wood for drawing the Wodrow volume to my attention.

12 See Whiteside, *Mathematical Papers*, viii, xx–xxvi; W. Jones (ed.), *Analysis per Quantitatem Series, Fluxiones, ac Differentias: cum Enumeratione Linearum Tertii Ordinis* (London, 1711); J.P. Bernard, T. Birch, J. Lockman et al., *A General Dictionary, Historical and Critical: In which a New and Accurate Translation of that of the Celebrated Mr. Bayle . . . is included; and interspersed with several thousand Lives never before published* (10 vols., London, 1734–41), vii, 776–802; S.P. Rigaud, *Historical Essay on the First Publication of Sir Isaac Newton's Principia* (Oxford, 1838). By the time Birch's expanded piece appeared in the *Biographia Britannica* in 1760, the papers had come into the possession of the Earl of Macclesfield, where they now remain.

13 *General Dictionary* (n. 12), vii, 776–802, esp. p. 793. See also ibid., iii, 556–7 and 558.

inclusion of his letter to Boyle of 1679 in Birch's 1744 edition of Boyle's *Works*, and the publication in 1756 of his 1692–93 letters to Richard Bentley. In 1757 Birch republished Newton's 'Hypothesis explaining the properties of light, initially read out at the Royal Society in December and January 1676, in volume 3 of his edition of the *History of the Royal Society*. With the increasing proliferation of material published in different locations, Samuel Horsley set out to produce an edition of Newton's works.[14]

Horsley sought not merely to publish well-known texts, but also any unknown manuscripts which would enhance Newton's reputation, and after some difficulty he was granted a brief stay at Hurstbourne in 1777 to inspect the papers. Horsley's list and William Mann Godschall's copy remain perhaps the final evidence regarding the initial ordering of Newton's papers, and as Whiteside points out, this is 'the first careful record we have of Newton's manuscripts and is invaluable as a check on the not wholly productive efforts of later cataloguers . . .'. Horsley found the papers divided up into individual parcels contained within drawers in a 'Cabinet' which he termed 'Cabinet A' and three other 'Letters' or 'Bureaux' marked 'B', 'C' and 'D'. In fact, the theological and chronological material, possibly rearranged by Sykes, was sprinkled in various drawers of the Cabinet and in Parcels 3 to 7 of one of the drawers of Bureau 'B'. In a copy of this list, four consecutive parcels in drawer 8 of the Cabinet relating to chronology are annotated with a star, as are three consecutive parcels in drawer 10 described respectively as 'Latin papers on the Jewish Temple', 'Theologiæ Gentilis Origines Philosophicæ' and 'Papers relating to the Chronology and Prophecies'.[15]

It is possible that Horsley had marked these as suitable for publication, though some of the same manuscripts (in language which recalled Pellet's ruthless dismissal of their publishable quality) were described as 'imperfect' or as a 'bundle of foul papers'. As it was, Horsley was at one point going to publish a version of the 'Geometria' which Newton composed in the early 1690s, although he thought better of it and it took another two centuries for the varieties of this remarkable work to go into print. Likewise, he suppressed the extraordinary 'Third letter to a friend', a follow-up letter to two previous accounts of corruptions in Scripture which, unlike the first two, left no doubts as to its unorthodox, antitrinitarian nature. However, his *Opera Omnia* largely satisfied the demands of eighteenth-century scholarship and indirectly prompted scholars such as Edward Gibbon to

14 The letters to Boyle and Bentley appeared in Birch (ed.), *The Works of the Honourable Robert Boyle* (5 vols., London, 1744), v, 70–3 and *Four Letters from Sir Isaac Newton to Doctor Bentley containing some arguments in proof of a Deity* (London, 1756); Newton's 'Hypothesis' was published in Birch's *History of the Royal Society* (4 vols., London, 1757), iii, 247–305 (and originally in *Philosophical Transactions*). Bryan Robinson reprinted the letter to Boyle with a commentary linking its doctrines to those of 'De natura acidorum' and the 'Queries' of the *Opticks* in *Sir Isaac Newton's Account of the Æther, with some additions by way of an appendix* (Dublin, 1745); prior to the initial printing of the letter, Robinson had published *A Dissertation on the Æther of Sir Isaac Newton* (Dublin, 1743).

15 Keynes MS 127A(4); Whiteside, *Mathematical Papers*, i, xxvi–xxvii, esp. xxvi n. 32. Horsley's edition appeared as *Isaaci Newtoni Opera quæ exstant Omnia. Commentariis illustrabat Samuel Horsley, LL.D., R.S.S.* (5 vols., London, 1779–85).

seek out individual manuscripts such as a version of 'Paradoxical questions concerning Athanasius' in preparation for his *Decline and Fall of the Roman Empire*. Moreover, it was with the authority as editor of these works that Horsley attempted unsuccessfully in 1784 to oust the virtuoso and incumbent President of the Royal Society, Joseph Banks, claiming that 'science herself had never been more signally insulted, than by the elevation of a mere *amateur* to occupy the chair once filled by Newton'. Not even Newton could exert sufficient posthumous influence to gain Horsley the Presidency and then the Bishopric which Banks's supporters accused him of desiring but he continued to point to Newton's Christianity in his attacks on the *philosophes* whose infidelity was, he argued, a primary cause of the French Revolution.[16]

Dismantling the Portsmouth Collection

Although his reputation changed considerably over the period, Newton remained by far the most important template for the role of the scientist. In the following nine decades, a few scholars were given permission to consult the papers at Hurstbourne while other sources also became available for study. For example, Joseph Edelston used previously unexamined material in Trinity College and in the Cambridge University and Royal Society libraries for the notes to 'A Synoptical View of Sir Isaac Newton's Life' in his 1850 edition of the correspondence between Newton and Roger Cotes. In particular, J.H. Monk, Francis Baily and David Brewster were fortunate that Henry Fellowes, the nephew of the third Earl of Portsmouth, was concerned about the deteriorating condition of the papers and granted them permission to consult the archive in connection with their respective editions of the lives of Richard Bentley, John Flamsteed and Newton. In fact, as Brewster admitted in the 'Preface' to his *Memoirs*, Fellowes himself had 'examined and arranged [a collection of manuscripts and correspondence] as peculiarly fitted to throw light on the Life and Discoveries of Sir Isaac', an undertaking which made his task much easier. Brewster only examined a small portion of the archive at Hurstbourne, and scholars did not have a chance to inspect the Hampshire manuscript corpus in its entirety until the 5th Earl, prompted by the patriotic and scientifically-interested Duke of Devonshire, made the remarkable offer of the 'scientific' part of the archive to the University of Cambridge in 1872. Portsmouth wrote that he wanted 'to advance the interests of science by placing these Papers

16 See CUL Add. MS 3963.11 fol. 112v and Gibbon to Joseph Warton, 22 June 1780, cited in F. Manuel, *Isaac Newton, Historian* (Cambridge, 1963), p. 255 n. 21. For the Banks episode see Anon, 'The Right Honourable Sir Joseph Banks', *Annual Biography and Obituary*, 5 (1821), 97–120, on p. 104. See also J. Heilbron, 'A mathematicians' mutiny, with morals', in P. Horwich (ed.), *World Changes: Thomas Kuhn and the Nature of Science* (London, 1993), pp. 81–130, esp. pp. 84–91 and 94, and J. Gascoigne, *Joseph Banks and the English Enlightenment: Useful Knowledge and Polite Culture* (Cambridge, 1994), pp. 62–3.

at the service of the University' but said that he would like to keep items relevant to Newton's personal life.[17]

As official delegates of the university, the task of examining the papers fell to the eminent Cambridge scientists George Stokes and John Couch Adams. When they realised that a thorough analysis of the papers would require more than the few days that had been employed by previous scholars, Portsmouth generously facilitated their examination of the archive by allowing them to take it back in its entirety to Cambridge. A passage from a letter sent by Lady Portsmouth to the Cambridge Vice-Chancellor in August 1872 gives a good indication of the perceived state of the archive. The reason for Portsmouth's allowing the papers to go to Cambridge was, she said, that the sorting of the 'confused' papers would have taken an 'indefinitely' long time, while at Cambridge the scholars could examine the papers 'with thoroughness and at leisure . . . Sir Isaac's papers appear never before now to have been carefully arranged or examined by any competent person so that in bundles marked "worthless" L[or]d P[ortsmou]th understood that the Professors thought they had found calculations of considerable interest & importance'. In the official syndicate appointed on 6th November 1872 to catalogue these papers, Adams and Stokes were joined by two other Cambridge scholars, G.D. Liveing, Professor of Chemistry, and H.R. Luard, the University Registry and a well-known medievalist, the former to look at the chemical and alchemical papers, and the latter to catalogue the personal and theological material. The very terms of their brief condemned the archive to be split according to the classification of late nineteenth-century science, and their findings were published in a *Catalogue* of the Portsmouth papers in 1888.[18]

In the 'Preface' to the catalogue the syndicate reported that having been appointed to 'examine, classify and divide' the entire Portsmouth collection, it had proved a 'lengthy and laborious business'. Many of the papers 'were found to be in great confusion – mathematical notes being often inserted in the middle of theological treatises, and even numbered leaves of MSS. having got out of order', and a large portion of the papers had been 'grievously damaged by fire and damp'.

17 See Whiteside, *Mathematical Papers*, i, xxvii–xxx; CUL Add. MS 2588. 494/5 and R. Yeo, 'Genius, method and morality: images of Newton in Britain, 1760–1860', *Science in Context*, 2 (1988), 257–84. Edelston's edition appeared as *Correspondence of Sir Isaac Newton and Professor Cotes, including Letters of Other Eminent Men, now published from the Originals in the Library of Trinity College, Cambridge* (Cambridge, 1850). Given the close proximity of the dates involved, it is probable that the Ekins family were asked to present their collection to New College by the Duke of Devonshire. Horsley managed to inspect all, and copy some of the Ekins archive in the preparation of his edition while Brewster saw it in 1855; see Ekins to Horsley, 26 January 1779 (cited in Manuel, *Historian* (n. 15), p. 255 n. 18) and Brewster, *Memoirs* (n. 9), ii, 341–2).

18 Whiteside, *Mathematical Papers*, i, xxx–xxxii; *A Catalogue of the Portsmouth Collection of Books and Papers written by or belonging to Sir Isaac Newton, the Scientific Part of which has been Presented by the Earl of Portsmouth to the University of Cambridge, drawn up by the Syndicate appointed 6th November 1872* (Cambridge, 1888). Liveing's most important published work was *Chemical Equilibrium the result of the Dissipation of Energy* (Cambridge, 1885). Luard published several works concerning the history of administration in Cambridge University, together with a six-volume *Catalogue of the Manuscripts Preserved in the University of Cambridge* (Cambridge, 1856–57) and a number of texts in the Rolls series.

Before going on to detail their own taxonomy of Newton's papers, the authors of the *Catalogue* dismissed Horsley's exertions, remarking that 'He left a few unimportant remarks on some of the papers, but he made no use of them in his edition.' Adams in particular took up about half the preface with his own account of Newton's researches into lunar theory, atmospheric refraction and the Solid of Least Resistance and, as Whiteside points out, the Appendix which gave a few extracts from Newton's papers was devoted entirely to examples from these three areas. These were topics in which Adams had more than a passing interest although he justified the space given to them by arguing that little manuscript material in these realms had so far been published.[19]

The 1888 Catalogue divided up the papers under the following headings: 'Mathematics', 'Chemistry', 'Chronology', 'History', 'Miscellaneous Papers, chiefly on theological subjects', 'Letters', 'Books', 'Miscellaneous papers', 'Correspondence', and minor sections dealing with material (such as the drafts of Conduitt's intended biography) connected with Newton. The rest of the Preface dealt with the fluxional controversy (Leibniz was castigated for ingratitude and possible theft), alchemy and theology. Although the alchemical papers appeared under the heading of chemistry, 'five parcels containing transcriptions from various alchemical authors in Newton's handwriting' and two other 'packets' containing alchemical notes and recipes were returned to Hurstbourne and condemned as being 'of very little interest in themselves'. Liveing concluded that 'He seems to have made transcripts from a variety of authors, and, if we may judge by the number of praxes of their content which he began and left unfinished, he seems to have striven in vain to trace a connected system in the processes described.'[20]

Along with the papers relating to the Mint, the items described as 'Chronology', 'History' and 'Theology' were also returned to Hurstbourne and the Syndicate concluded that these manuscripts 'cannot be considered of any great value', an unintended slight to the Portsmouth family given that in 1872 the 5th Earl had offered the University the material relating to the so-called 'Father Francis' affair (in which Newton had argued that King James II's attempt to impose a Catholic M.A. upon Sidney Sussex College was illegal). The 'Chronological' material dated from the last two decades of Newton's life, while the 'Historical' section was almost entirely composed of Newton's defence of Sidney Sussex. Luard ended his classification of this section with three items on 'Egyptian mythology', the 'Original of Monarchies' and some 'Antiquarian fragments'. The difficulties associated with distinguishing between chronological and theological subjects is further exemplified by the material which appeared in Section V under the heading of 'Miscellaneous papers, chiefly on theological subjects'. This contained chronological material as well as versions of his interpretation of the apocalypse and 'Paradoxical questions concerning Athanasius', although it also embraced the notes for and drafts of the 'Theologiæ Gentilis Origines Philosophicæ'. Left with

19 *Catalogue* (n. 18), pp. ix–xi and xxi–xxx; Whiteside, *Mathematical Papers*, i, xxxiii n. 48. Horsley's annotations are now in CUL Add. MS 4005.

20 *Catalogue* (n. 17), pp. 11, xix.

a motley collection of 'non-scientific' writings, Luard struggled unsuccessfully to come to terms with the purpose of any of this work and with the mentality underlying the creation of so many drafts, and he remarked that 'A great portion of Newton's later years must have been spent in writing and rewriting his ideas in certain points of Theology and Chronology':

> Much is written out, as if prepared for the press, apparently from the mere love of writing. His power of writing a beautiful hand was evidently a snare to him. And his fastidiousness as to the expression of what he wrote comes out very curiously in these papers . . . The four elaborately bound volumes containing 'the Chronology of Ancient Kingdoms Amended', the Chronicle to the Conquest of Persia by Alexander, Observations on the Prophecies, and the treatise 'De Mundi Systemate', are very remarkable specimens in their author's care in writing out his works, and of his beautiful handwriting.

No-one made use of the truncated Portsmouth collection until Lady Portsmouth noble-mindedly sent the papers to the British Museum in the late 1920s for the benefit of Louis Trenchard More, then engaged on his biography of Newton. The full extent of Newton's alchemical and theological work only came to light with the publication of Sotheby's comprehensive sale catalogue of the Portsmouth Papers in 1936.[21]

Peter Spargo has shown that the remainder of the Portsmouth archive which was sold at Sotheby's was heavily undervalued because of the lure of the competing sale of Henry Oppenheimer's collection of Impressionist art at Christie's. The manuscripts arrived in a large metal trunk at Sotheby's three or four months before the sale actually took place on 13 and 14 July, and the task of organising the papers and producing what was ultimately an excellent catalogue fell to J.C. Taylor, the Chief Cataloguer in the Book and Manuscripts Department. The sale attracted little major interest until it was over, when John Maynard Keynes and Abraham Yahuda went about frantically buying up manuscripts from dealers in order to supplement their respective collections.[22] The Keynes collection was bequeathed to King's College on Keynes' death in 1946, and in his magnificent work on Newton as historian, Frank Manuel made substantial use of it. Manuel published a transcription of the 'Original of Monarchies', and drew from material in the Royal Society, Cambridge University and British Libraries, as well as from other papers bought

[21] Ibid., pp. xix–xx, 25–31; L.T. More, *Isaac Newton, A Biography* (Chicago, 1936), p. ix. The volumes mentioned by Luard were catalogued under section VII, 'Books'. Portsmouth's offer to 'hand over' the 'bundle of papers' relating to the Father Francis affair is now CUL Add. MS 2588 (495). Papers relating to Newton's time at the Mint were bought by Lord Wakefield from the dealer Gabriel Wells and immediately given to the Royal Mint. They are now in the Public Record Office. The only major study of their content is J. Craig, *Newton at the Mint* (Cambridge, 1946).

[22] [J. Taylor], *Catalogue of the Newton Papers Sold by the Order of The Viscount Lymington* (London 1936); P. Spargo, 'Sotheby's, Keynes and Yahuda – the 1936 sale of Newton's manuscripts', in P.M. Harman and A. Shapiro (eds.), *The Investigation of Difficult Things. Essays on Newton and the Exact Sciences* (Cambridge, 1995), pp. 115–34, on pp. 116–17. Lymington was also about to become deeply implicated in neo-fascist politics: see R. Griffiths, *Fellow Travellers of the Right. British Enthusiasts for Nazi Germany 1933–39* (Oxford, 1983), pp. 317–29. I owe this reference to David Edgerton.

from the 1936 sale in institutions such as the Royal Mint, and the King's College, Pierpont Morgan, Babson Institute, and Trinity College Libraries. Unable in the early 1960s to access the treasures of the Yahuda collection, unaware of its existence as an extended collection, and left almost entirely with the far more turgid work of the Newton of the early eighteenth century, Manuel immersed himself in the New College archive although he admitted that Newton's later historical endeavours 'cannot in good faith be recommended for their liveliness to a reader of the second half of the twentieth century'.[23]

Within three years of the appearance of Manuel's book the Yahuda collection, which contained the majority of Newton's seventeenth-century theological papers, arrived in Jerusalem and was made available to scholars. Manuel himself was one of the first to avail himself of the opportunity of examining this collection, culminating in the publication of his Fremantle lectures which were based on the archive. All major Newton material was now in the public domain, and it was possible to compile a comprehensive picture of Newton's intellectual life. Taking into account almost all the known Newton manuscripts and a wealth of other associated sources, R.S. Westfall has been the only historian in the modern era to claim to have read virtually all of this Newtoniana in preparation for his biography. Although he admitted that Newton the man had receded from him the more he had studied him, and had become 'wholly other . . . a man not finally reducible to the criteria by which we comprehend our fellow beings', he did not see Newton as irreconcilably torn between the occult and scientific, and made a substantial effort to view his work in the context of the totality of Newton's intellectual interests. Nevertheless, he indicated that since Newton held our interest 'only because he was a scientist of transcendent importance', he tended to think of his own work 'as a scientific biography, that is, a biography in which Newton's scientific career furnishes the central theme'.[24]

Westfall's justification for producing a scientifically-oriented biography of Newton is significant. Clearly, the very survival of his papers has depended on the degree of reverence that has been paid to his scientific achievement, and yet once that is recognised, there seems to be no need to continue to ignore or downplay what Newton *himself* might have taken to be the most significant part of his intellectual pursuits. Indeed, it is arguable that Newton's theological papers – despite their extreme unorthodoxy – are important in their own right for understanding wider contemporary religious issues. With this in mind, a case can be

23 Manuel, *Historian* (n. 15), p. vii and esp. p. 254 n. 14: 'Repetitions are so frequent as to make it doubtful that the remaining pieces [i.e. most of the lots which were bought by Yahuda] would alter the general propositions set forth in this book.' 'The Original of Monarchies' was lot 295 at the 1936 sale and is now Keynes MS 146; Manuel's transcription with scholarly apparatus is in ibid., pp. 195–221 and 303–12. For whatever reason, this lot was particularly desired by Lymington himself, who lamented to Keynes in a letter of 8 September 1936 that he had carelessly overlooked it 'until just after the hammer fell' and that therefore he did not deserve it; cf. Spargo, 'Sotheby's' (n. 21), pp. 121–2.

24 Spargo, 'Sotheby's' (n. 21), pp. 125–8; Westfall, *Never at Rest: A Biography of Isaac Newton* (Cambridge, 1980), pp. x–xi. For the extent to which Westfall recognised that his subject had been painted in his own image see his 'Newton and his biographer', in S.H. Baron and Carl Pletsch (eds.), *Introspection in Biography: The Biographer's Quest for Self-Awareness* (Hillside, N.J., 1985), pp. 175–89.

made for privileging Newton's theological interests above any others; it should be recalled that it was precisely these concerns that most fascinated his eighteenth-century followers in the period immediately after his death.

Nevertheless, the part of the Portsmouth Collection examined by Stokes and Adams has inevitably provided the most significant resource for Newton scholarship, and has generally been considered to be the most important portion of the archive as a whole. Almost immediately after the collection was made accessible in Cambridge University Library in 1888, W.W. Rouse Ball published a history of the development of Newton's *Principia*,[25] but interest in Newton 'took off' with the professionalisation of history and philosophy of science in the 1950s, as constituted in the once mighty 'Newton industry'. Thanks to the pioneering work of a number of scholars in the last fifty years, the existence of such a resource has transformed the historical enterprise in such a way that historians have for some time been able to document minute changes in Newton's development in mathematics, optics, mechanics and, to a lesser extent, chemistry. The manuscripts have taken on a significance for understanding Newton's thought which is greater than that of his published texts, and, having moved away from an obsession with the published text as the *telos* of incomplete and hence inferior draft versions, the papers which in the eighteenth and nineteenth centuries had been deemed 'incomplete' or 'not fit to be printed' have now been rescued from their obscurity.[26]

At the beginning of the decade which celebrated the tercentenary of Newton's *annus mirabilis*, Rupert and Marie Boas Hall published some of the most interesting of the manuscripts from the Cambridge University holdings, while in the most comprehensive of these efforts, Tom Whiteside devoted eight large volumes (replete with omniscient footnotes) to Newton's mathematical progress. Paying little attention to the store of papers which testified to Newton's less palatable interests, he concluded the 'General Introduction' to volume 8 with the hope that his edition would 'drive out the entrenched myths about their provenance and character which ought to be no longer tenable, and strangle at mis-birth the yet more fanciful notions which float in the train of currently trundling band-wagons'. The stature of Whiteside's achievement is shown by the fact that virtually the only work now being done in what is left of the Newton 'industry' revolves around his edition of the mathematical papers, and scholars cite page numbers from volumes of the *Mathematical Papers* in much the same way that Whiteside once did individual folios from the Cambridge collection.[27]

25 W.W. Rouse Ball, *An Essay on Newton's 'Principia'* (London, 1893).

26 Newton's chemical notebook, CUL Add. MS 3975, contains dated entries as late as 1693 and was studied in a pioneering article by A.R. and M.B. Hall, 'Newton's chemical experiments', *Archives internationales d'histoire des sciences*, 11 (1958), 113–52.

27 Hall and Hall, *Scientific Papers* (n. 2); Whiteside, *Mathematical Papers*, viii, xxvii–viii. Whiteside also published facsimiles of unpublished material relevant to the early work on the *Principia* in the mid-1680s in *The Preliminary Manuscripts for Isaac Newton's 1687 Principia* 1684–1686 (Cambridge, 1989). For recent work on Newton's studies in mechanics, see in particular J. Bruce Brackenridge, 'The Locke/Newton manuscripts revisited: conjugates, curvatures and conjectures', *Archives internationales d'histoire des sciences*, 43 (1993), 280–92, and M. Nauenberg, 'Newton's early computational method for dynamics', *Archive for History of Exact Sciences*, 46 (1994), 221–52.

Before the completion of the *Mathematical Papers*, Newton's private researches in optics had been much better documented than any other area of his work. In the first volume of his edition of Newton's optical letters and papers, Alan Shapiro promised to give a definitive account of Newton's optical work, 'so that the evolution and formulation of Newton's ideas may be followed through early versions, drafts, and revisions in the papers'. This would differ from eighteenth-century expositions and editions of his works, 'the primary aim of which was to present Newton's definitive scientific position'. Future volumes were to reproduce the bulk of the still unpublished optical and related researches, many of which branched out in the early eighteenth century to link topics of chemistry, light, life and electricity. Shapiro also argued that these tomes should include drafts of letters (whether these letters were private or later published in the *Philosophical Transactions*) since the editors of Newton's correspondence were neither consistent nor exhaustive in noting variant readings or expunged passages, either in drafts or in the letters which were actually sent. Nevertheless, while perhaps not strictly part of his 'archive', the *Newton Correspondence* has been the most widely read and cited of all of the publications of Newton's unpublished corpus.[28]

As a corrective to the occasionally one-sided stress on the more obviously 'unscientific' areas of Newton's work, few could argue after these and many other publications that Newton was not thoroughly immersed in and a master of contemporary optical and mathematical work, or that these did not continue to be immensely important areas of his intellectual life. Despite the vast amount of writing in the hand of his amanuensis (Humphrey Newton) between 1684 and 1691, there is still no hard evidence to dispute Whiteside's assertion that Newton took no time off from the twenty-three months' labour it required to compose the *Principia*. Nevertheless, statements about the general primacy *for Newton* of these concerns above say, theology, are highly dubious when one considers the evidence for the extraordinary intellectual and emotional toil expended upon the latter from the early 1670s. Nor are areas of enquiry such as optics always easily distinguishable from other research which Newton himself believed to be relevant. In his work on the nature of light, for example, his experimental researches were often related to or drew upon other intellectual endeavours, amongst which must be also included alchemy. Finally, even within what is conventionally taken to be the 'scientific' part of Newton's enquiries, the question of how to reconcile the very different methodologies and contents of the *Principia* and the *Opticks*, in particular its 'Queries', has long been a problem for attempts to posit a single 'essence' to Newton's science.[29]

[28] Shapiro, *Optical Lectures* (n. 2), i, xi–xii; most of the relevant material on his optical reseaches is now in CUL Add. MS 3970. J.E. McGuire's and M. Tamny's edition of Newton's Trinity Notebook (now CUL Add. MS 3996), *'Certain Philosophical Questions': Newton's Trinity Notebook* (Cambridge, 1984), is an important addition to the total of Newton writings now in print, and sheds a great deal of light on his entry into optics. This text also includes relevant sections on Newton's experiments on colours from his slightly later experimental notebook CUL Add. MS 3975.

[29] Dobbs, *Janus Faces* (n. 2), passim. For comments on the increasingly authoritative status of unpublished archives in general, see Schaffer and Shapin, 'Making Newton' (n. 4), p. 12, and pp. 15–23 for the problem of the 'coherence' of Newton's science.

Projects and publication

Having throughout this essay referred to Newton's 'archive', it should be stressed that this entity is obviously somewhat nebulous. The sheer material bulk of the numerous papers stained with ink from Newton's pen are a testament to the writing effort expended during his lifetime, but quantity *per se* can say little about the nature of these efforts. On the one hand, the manuscript collection does not include many comments made about Newton both during and after his lifetime and it obviously excludes the many works published throughout the eighteenth century which defended or attacked his views. As with correspondence written to Newton, all these should be taken into consideration if one wishes to understand his work in the light of wider 'contexts'. On the other hand, the Newton archive is understood to include the autograph both of amanuenses and of other unidentifiable persons, and it is often difficult or even impossible to tell whether notes or papers represent Newton's 'original' thoughts or 'merely' his adaptations of different texts, or even, in the cases of theology and alchemy, whether they are 'his' thoughts at all.[30]

Such problems of authorship are related to the more general question of whether private thoughts are to be treated as the 'real' beliefs of individuals – what Newton 'really' thought – or as preliminary and incomplete musings. As we have already seen, in the case of Newton it is all too easy to argue that drafts or suppressed passages represent 'inferior' versions of published texts. It is equally plausible to hold – as many of his contemporaries did – that public pronouncements on, say, gravitation, were a cover for the real opinions whose publication would engage him in all kinds of unwanted controversy. The perceptive reader could find 'hints' in Newton's published texts for beliefs which were expressed in private manuscript, while those without access to his personal papers could authoritatively be dismissed for their 'misunderstanding' of his 'real' meaning. Consideration of these and similar points entails that from the perspective of historical balance, a fetishistic concern with manuscript sources is as unhealthy as an obsession with published texts, 'print culture' and the 'public sphere'.[31]

Although many manuscripts disappeared for various reasons during Newton's lifetime, his dog-earing and annotating of books, and his notes and drafts have left a vivid testament to his working practices.[32] In garnering information for use in theological and historical areas, he made substantial use of repositories such as the compendia of Comes and Cartari, and of grandiose projects such as Cudworth's

30 W. Newman, 'Newton's "Clavis" as Starkey's "Key" ', *Isis*, 78 (1987), 564–74. With the availability of microfilm, archivists are increasingly unwilling to allow access to the manuscripts themselves.

31 Cf. Schaffer and Shapin, 'Making Newton' (n. 4), pp. 6–7, 12 and 30–5.

32 For a good account of Newton's use of books, see Harrison, *Library* (n. 6), pp. 1–27, and cf. K. Figala, J. Harrison and U. Petzold, '*De Scriptoribus Chemicis*: sources for the establishment of Isaac Newton's (al)chemical library', in Harman and Shapiro, *The investigation of difficult things* (n. 21), pp. 135–79. Brewster cited Samuel Crell's opinion that Newton burned a number of manuscripts before he died; *Memoirs* (n. 9), ii, 390, and cf. Whiteside, *Mathematical Papers*, i, x. As an anti-Trinitarian confidant of the ageing hero, Crell's testimony of Newton's actions should be taken seriously.

True Intellectual System, though his debt to these sources was almost never recorded in his more completed work. In all realms of his scholarly activity, certain favoured manuscripts or essays were repeatedly consulted, occasionally many decades later, and when the need arose he was prepared to write on practically any scrap of paper which was to hand. As Frank Manuel noted in his psychoanalytic biography of Newton, many letters – such as the only surviving example from his mother – are ruthlessly and (for Manuel) surrealistically scrawled over with notes on subjects such as the calculus, the games played by the Devil in the third or fourth centuries, or even with the scientific achievements of Chiron the Centaur.[33]

As we have seen, historians have long pondered the function of the continual rewriting of drafts with which Newton persisted for half a century. Beyond noting the fact that he was not writing for publication or remarking on the dissatisfaction which he obviously felt about the many versions of his work, I have no special insight into his 'intentions'. Luard's own inability to divine any purpose to Newton's constant redrafting was linked to his suggestion that his work was written 'as if prepared for the press', but although this description of the state of many of the essays is accurate, Newton notoriously had little taste for appearing in print except for a brief period in 1671 and early 1672 when he threatened to publish material on optics and infinite series, and to edit Kinckhuysen's *Algebra*.[34] Once he had effectively severed his relations with the community of natural philosophers in London, the publication of the first edition of *Principia* was apparently only achieved after a great deal of public unwillingness on the author's behalf. Even after he had achieved prominence in the metropolis in the last few years of the seventeenth century, he remained guarded about publicising untimely thoughts in alchemy and theology, and his views on natural philosophy were notoriously propounded as hypothetical 'Queries' in the 1704 *Opticks* and subsequent editions.[35]

In considering this issue it is perhaps better to think of Newton's *audience*,

33 Manuel, *Portrait* (n. 3), pp. 14–16, esp. p. 15.

34 The first volume of Newton's *Optical Papers* details the transformation of a version of Newton's original set of Lucasian lectures (the 'Lectiones Opticae', now CUL Add. MS 4002) into an expanded and publishable form ('Opticae, Pars 1a/2a', now CUL Dd.9.67). Newton worked on the second of these over the autumn and winter of 1671/2, and the kernel of his doctrine was published in *Philosophical Transactions* in February 1671/2. Isaac Barrow told John Collins of Newton's plans for publishing '20 Dioptrick lectures' (which Barrow 'reckon[ed] one of the greatest performances of Ingenuity this age affoarded') along with his 'De methodis seriorum et fluxionum', but nothing came of it after Newton tired of the controversies engendered by his new theory of light and colours. See Shapiro, *Optical Papers* (n. 2), i, 15–19, esp. p. 15 n. 51; *Newton Correspondence*, i, 30–1 (for Kinckhuysen), 79–107 and 146, and Whiteside, *Mathematical Papers*, iii, 3–9. Newton did achieve authorship with his edition of Bernard Varenius's *Geographia Universalis* (Cambridge, 1672); see *Newton Correspondence*, i, 23 n. 14, 146 and 161–2.

35 In the Preface to the first edition of the *Principia* he claimed that it was through the 'solicitations' of Halley that the work was published after the latter had 'continually pressed' him into sending a demonstration of the shape of celestial orbits to the Royal Society, and in *Opticks* he remarked regarding the controversies that had surrounded his work in the 1670s: 'To avoid being engaged in Disputes about these Matters, I have hitherto delayed the printing, and should still have delayed it, had not the Importunity of Friends prevailed upon me'; see *Principia*, ed. F. Cajori (2 vols., London, 1962), i, xviii

whose size and competency ranged in a continuum from himself to the large and anonymous aggregate of readers of his printed *oeuvres*. Although he professed to despise the printed text as a forum for disseminating important information, he just as frequently claimed that he was not averse to releasing his papers to a select group of 'friends' who could discuss his work in peace without disputing his veracity or competency. This of course was particularly relevant when the subject matter of his alchemy and theology would have been considered beyond the pale by the majority of his contemporaries, but he also demanded a restricted audience for his work on light and colours in the 1670s. It was these real or imaginary elite peer groups for whom he wrote in mathematics, alchemy, theology and natural philosophy and in nearly all the cases in these fields where we have virtually 'finished' texts, there is little indication that Newton had any plans for their broad distribution, and still less for them being printed. Unfortunately, given what could have been at best a tiny circulation, we have often almost no evidence about the real audience for these papers.[36]

In many cases, the subject of such extended texts would be transformed in further drafts or even altered as his view of the purpose of the exercise changed. In terms of presenting such manuscripts to a modern public this clearly creates something of a problem. In many cases, an editor of Newton's work is faced by the lack of any privileged *published* text against which one can view others as drafts or variants, despite the fact that Luard was correct about the relatively finished *appearance* of much of this work.[37] However, this turns out to be deceptive since on many occasions Newton neared completion of one project only to drop it as the text apparently drew to a close, and the fact that he rarely if ever considered a work 'finished' is evidenced by the numerous drafts and 'versions' of even his most famous creations. In this vein, it is significant that he planned very different versions of the *Principia* and *Opticks* from those which appeared in the respective published editions. For example, after the publication of *Principia Mathematica* in 1687 (and roughly the same time that he began revisions of the 'Fundamentum Opticae', a preliminary version of Book One of the *Opticks*), his private views on the nature of a second edition became known to a select group

and *Opticks, or A Treatise of the Reflections, Refractions, Inflections & Colours of Light* (New York, 1979), p. cxxi.

36 For one of many attempts to stipulate a restricted, conversational sphere as the proper forum for discussing his work see Newton to Hooke, 5 February 1676; *Newton Correspondence*, i, 416. For indications of possible sources of and an audience for Newton's alchemical writings, see K. Figala and U. Petzold, 'Alchemy in the Newtonian circle: personal acquaintances and the problem of the late phase of Isaac Newton's alchemy', in J.V. Field and F.A.J.L. James (eds.), *Renaissance and Revolution* (Cambridge, 1993), pp. 173–92.

37 See M. Hunter, 'How to edit a seventeenth century manuscript: principles and practice', *The Seventeenth Century*, 10 (1995), 277–310, esp. pp. 290–1. In view of Newton's continual redrafting of his work, I am less confident than Hunter about the possibility of reconstructing a single definitive text (cf. ibid., p. 281). With the exception of the two letters on corruptions of Scripture which he gave permission to print anonymously (but which he later suppressed), it is highly unlikely that the printing of his work ever loomed large in Newton's consciousness. On the issue of individual intentions which this raises see the detailed discussion in G.T. Tanselle, 'Textual instability and editorial idealism', *Studies in Bibliography* (1996), 1–60, esp. pp. 25–41 and 53–8.

including Christiaan Huygens, Gottfried Wilhelm Leibniz, Fatio de Duillier and David Gregory, and plans were made to produce a new, corrected edition, with snippets from ancient philosophies proposed as new scholia (the 'Classical Scholia'). It is thanks to Gregory's visits to Cambridge in 1694 that we have what is perhaps the first major independent account of the state of his researches.[38]

Newton undertook a colossal revision of all the areas of his intellectual activity in the early 1690s and he embarked upon a reorganisation of many of his optical, mathematical, alchemical and theological ideas. In the case of the *Principia*, he recaptured the confidence of the mid-1680s when he had planned to publish a version of the great work in which it was demonstrated that almost every recorded society, from the ancient world to modern pagans, had based the central rituals of their culture around the truth of the heliocentric cosmos. The ancients had known of the Newtonian philosophy but had disguised it so as not to confuse the vulgar, although the vestal religion depicted heliocentrism by virtue of its worship around a central fire. Although Fatio de Duillier heard of Newton's interest in this over two years earlier, David Gregory's notes from May and July 1694 represent the first detailed account of Newton's work in this area. However, the initial extended treatment of this topic is presumably that constituted by the jumble of notes in preparation for the *Principia* which are now in Jerusalem. The draft notes for a long essay on the ancients' theologico-philosophy, the 'Theologiæ Gentilis Origines Philosophicæ' (in the hand of Newton's amanuensis from the mid-1680s), were expanded into a number of different areas including the 'Original of Religions' of c. 1693–94 and then the 'Original of Monarchies' which dates from 1701–2. Gregory's attention was also drawn to an extraordinary essay entitled 'Geometria' (of whose existence Newton had informed him in London in December 1691) which concerned the 'long neglected genre of discovery' of the ancients in geometry. As with the fate that befell his plan to compose an all-embracing *Principia* in 1685, his initial sweeping goal of restoring the lost knowledge of the ancients disappeared as the work progressed and, as Whiteside remarks, it 'became a thorough-going contemporary treatise on curves and their properties'.[39]

For various reasons, it has been the fate of Newton's work on the origin of the

[38] For Gregory's visits to Cambridge see *Newton Correspondence*, iii, 327–40 and 384–9; cf. also W.G. Hiscock (ed.), *David Gregory, Isaac Newton and their Circle: Extracts from David Gregory's Memoranda 1677–1708* (Oxford, 1937), p. 4. Amongst those who had access to Newton's mathematical papers were John Craig, David Gregory, Joseph Raphson, Edmond Halley, Fatio, William Jones, John Keill, Henry Pemberton, Abraham de Moivre, Niklaus (Nicholas) Bernoulli and possibly John Wallis; Whiteside, *Mathematical Papers*, i, xvii.

[39] *Newton Correspondence*, iii, 191, 327–40 and 385–6 (for the 'Geometria'); Whiteside, *Mathematical Papers*, vii, 185–561 (for relevant texts), esp. pp. 186, 193; R.S. Westfall, 'Isaac Newton's *Theologiae Gentilis Origines Philosophicae*', in W.W. Wagar (ed.), *The Secular Mind* (New York, 1982), pp. 15–34. The dating for the 'Original Of Monarchies' is independently supported by the researches in to the watermarks of Newton's paper by Alan Shapiro; see his 'Beyond the Dating game: watermark clusters and the composition of Newton's *Opticks*', in Harman and Shapiro, *The Investigation of Difficult Things* (n. 21), pp. 181–227, on p. 197. Shapiro argued that the value of this exercise was extremely limited, although it is practically the only independent evidence there is for dating manuscripts. The relevant Jerusalem documents are now Yahuda MSS 13.3, 16.1, 16.2, 17.2, 17.3 and 41.

true religion and on the Classical Scholia to have become separated. Evidence from the correspondence of Fatio de Duillier suggests that Newton was planning to add classical references to Propositions 4 to 9 of Book 3 of the *Principia* in 1691–92; Newton gave Gregory copies of completed Classical Scholia in 1694 and may have been working on another version of them at the same time. At this point, Gregory saw or was informed about Newton's work on the origin of religion and nations. He mentioned that Newton would show that the ancients 'observed the gravitation of all bodies towards all' (a fact referred to by Plutarch and Galileo) and that 'the Egyptians and others taught the Copernican system, as he shows from their religion and hieroglyphs and images of the Gods'.[40] In important respects, Newton did not believe that the 'Classical Scholia' and his work on the corruption of the ancient religion were distinct, although scholars have not yet succeeded in connecting these additions to other areas where he wrote at length about the *prisca sapientia* tradition. For example, Paulo Casini published an important article on the 'Classical Scholia' without mentioning the existence of either the drafts or the fuller versions of the 'Philosophical Origins of the Gentile Theology'. Whatever the real divisions between disciplinary approaches which informed Newton's intellectual work, the dissemination of certain of his works to others and the subsequent classifications of his works in these areas has often caused modern scholars to lose sight of possible connections between regions of his work.[41]

Separate Spheres

Newton often drew from one area of his work to shed light on another, and occasionally he experimented with the use of certain techniques of persuasion or styles of presentation in a different field. For example, in one of his treatises on the prophecies, a disquisition on the meaning of the colour of the gemstones in Aaron's breastplate suddenly draws from his work on 'active principles', while a number of his early attempts to write a treatise on the apocalypse utilise the propositional form used in his mathematical work. I have already pointed to the ways in which Newton's interests in alchemy and chemistry were occasionally linked to his work on light, and I have argued that for Newton there was a fundamental unity of thought which connected the *Principia* to the study of ancient religion and natural philosophy. In one sense this is hardly surprising, since it was a feature of natural philosophy that part of its purpose was the glorification of God through the appreciation of His creation of the natural world.[42]

However, it is also vital to note that Newton himself rigorously divided up his own endeavours into two broad spheres. On the one hand, he maintained that his

40 *Newton Correspondence*, iii, 191 and 384 (c. July 1694); McGuire and Rattansi, 'Pipes of Pan' (n. 2), passim. The reference to the origin of nations is probably to Yahuda MS 41, 'The Original of Religions', although Gregory's notes may be only from discussion with Newton.

41 Yahuda MS 13.3; P. Casini, 'Newton: the Classical Scholia', *History of Science*, 22 (1984), 1–58.

42 For the passage on 'active principles' see Yahuda MS 9.2, fols. 7–7v (from the late 1680s).

theological work rested on faith and could not be demonstrated with mathematical reasoning and certainty to those who lacked it, and he argued that those who thought otherwise, such as Henry More, were perverse. On the other hand, certain methodologies were utterly inappropriate for doing natural philosophy, and he famously complained about the descent of natural philosophy into the litigious style redolent of judicial hermeneutics. For understanding both sacred and secular history, the moral probity of various authors, the 'good sense' of the text (if it countered the obscure language of the Trinity it was probably older and thus authentic), and the empirically validated authenticity of writings was crucial. What mattered in natural philosophy was the performance of experiments and the grounding of these in the mathematical sciences which were capable of being brought to the highest degree of certainty. Attempts such as those by Robert Boyle to base natural philosophy on the testimony given by witnesses were anathema to Newton, as were what he took to be the litigious and disputacious approaches to natural philosophy both of his Jesuit commentators in the 1670s and of Robert Hooke in the 1680s.[43]

In both of these spheres Newton believed he was one of a select group of people who had arrived at a superior understanding of nature or of Scripture, and it was his task to disseminate this truth. As such, one should perhaps consider the reflexive aspect of his work, that is, the sense in which he was implicitly writing of his own situation when he wrote of the role of the ancient priests of nature and of their problems in communicating truth to a particular audience. For such men knowledge of the highest religious and natural philosophical truths were equally necessary as a basis for their authority and I would suggest that Newton saw himself as a modern version of an ancient priest, morally entitled to be the guardian of the elite knowledge which he had attained in a number of different pursuits through honest toil. In virtually all of his writings, Newton repeatedly said that his audience was the select and worthy few. It is the connection between religious faith and natural knowledge which Newton recognised in the virtuous person of the ancient priest of nature – a fact which implies that a consideration of his archive *as a whole* has significant consequences for investigating any one area of his activity.[44]

43 For the comment on the inapplicability of mathematical levels of certainty to theology see R. Iliffe, ' "Making a shew": apocalyptic hermeneutics and Christian idolatry in the work of Isaac Newton and Henry More', in J.E. Force and R. Popkin (eds.), *The Books of Nature and Scripture* (Dordrecht: Kluwer, 1994), pp. 55–88, esp. p. 87 n. 62.

44 For reflexivity, see R. Iliffe, ' "Is he like other men?" Newton, the *Principia* and the author as idol', in G. Maclean (ed.), *Culture and Society in the Stuart Restoration* (Cambridge, 1994), pp. 159–76.

10

'A chaos of jottings that I do not have the leisure to arrange and mark with headings': Leibniz's manuscript papers and their repository*

James G. O'Hara

Introduction: Leibniz's archive

On 24 February 1693, in the early stages of their developing correspondence, the French mathematician and champion of the differential and integral calculus Guillaume François de L'Hospital wrote the following lines to Leibniz:

> It is a real pleasure for me to exchange ideas with a man of your learning. I have known for a long time that you are a universal man, theology, history, the rights of princes, the investigation of mines, etc., being your everyday occupation, and you have scarcely a moment to employ in mathematics and natural philosophy. Nevertheless, the great discoveries that you have made in these subjects, and are still making every day, are sufficient to make one aware of what things of this kind you are capable, and one cannot complain too much that you have so little leisure to think about them.[1]

L'Hospital, in these words of tribute to the polymath whose interests, as he saw it, ranged from theology, through history, jurisprudence, mining engineering, mathematics and physics, was inadvertently highlighting the difficulty the historian and archivist faces today in investigating the archive of the great early-Enlightenment philosopher and universal scholar, Gottfried Wilhelm Leibniz. From the latter's reply we learn of the multifarious duties of Leibniz that prevented more intensive mathematical endeavours, including an amount of unanswered correspondence that required considerable effort and investigation: his letters to

* I am grateful to my colleague Heinz-Jürgen Hess for support and helpful suggestions in the preparation and writing of this paper.

1 'Je me fais un vrai plaisir d'avoir quelque commerce avec une personne de vôtre erudition. Il y a longtemps que je sçais que vous étes universel, la theologie, l'histoire, les droits des princes, la recherche des mines etc. sont vôtre occupation ordinaire et à peine avez vous quelques momens pour les employer aux mathematiques et à la physique; cependent les grandes découvertes que vous y avez faites et que vous y faites encore tous les jours font assez connoistre de quoi vous étes capable en ce genre, et on ne sauroit trop se plaindre de ce que vous avez si peu de loisir à y penser.' G.W. Leibniz, *Mathematische Schriften*, ed. C.I. Gerhardt (7 vols., Berlin, later Halle, 1849–63; reprint Hildesheim, 1962), ii, 223–7.

like L'Hospital and Christiaan Huygens did not seldom have the length and substance of a scientific paper. There were in addition other duties at the court and obligations to friends, the study of new publications and the need for the civil servant and privy counsellor to keep abreast of current affairs. In moments of inspiration he would record ideas on paper, but his duties would divert his attention and he was forced later to search anew through a chaos of papers, which he could never find time to sort and arrange. The exact words of Leibniz's reply to L'Hospital's compliment in mid-March 1693 are as follows:

> It is not the universality of knowledge that you, sir, have attributed to me by the pure grace of your generosity that prevents me from satisfying my inclination for mathematics, but countless small matters that divert me. I believe that at present I have more than thirty letters waiting for a reply, in which it is always necessary to write something more than just compliments. And beyond the duties of my offices, one owes time to the court and to one's friends. Moreover, thoughts sometimes occur to me that I am pleased to preserve, new books need to be looked at, it is necessary to have some information on current affairs. And apart from the learned, if those who know me knew that on top of all this I also indulged in algebra, they would find it strange. After having done something, I forget it almost entirely within a few months, and rather than searching for it amid a chaos of jottings that I do not have the leisure to arrange and mark with headings, I am obliged to do the work all over again.[2]

The condition of Leibniz's papers during his mature and later years was a result of a number of factors unique to him. In addition to the universality of interests hinted at by L'Hospital, the state of his papers was a function of his very special biography. Throughout his life Leibniz was a traveller whose fixed abode was to become Hanover for the last forty years of his life (1676–1716). Legal studies at the university of his native city, Leipzig, and doctoral qualification at the University of Altdorf were followed by early employment in the service of the Elector of Mainz that culminated in a diplomatic mission to Paris. This led to a four-year sojourn in the French capital (1672–76) during which Leibniz found a mentor, in the person of Christiaan Huygens, in the course of his development as a mathematician. Journeys to England and Holland preceded his arrival to take up an appointment at the court of Hanover in 1676. His first major commission in Hanover was as engineering inspector in the Harz mining district, to which he

2 'Ce n'est pas cette universalité de connoissance, que vous m'attribués, Monsieur, par une pure grace de vostre liberalité, qui m'empeche de satisfaire à mon inclination pour les Mathematiques; mais une infinité de petites choses qui me detournent. Je crois d'avoir maintenant plus de 30 lettres qui attendent reponse où if faut tousjours dire quelque autre chose que des complimens. Et outre les devoirs de mes charges on doit du temps à la cour et à ses amis; de plus ils me viennent quelques fois des pensées que je suis bien aise de conserver; il faut voir les livres nouveaux; il est necessaire d'avoir quelque information des affaires courantes. Et excepté les sçavans si ceux qui me connoissent sçavoient qu'avec cela je m'amusois encor à l'Algebre, ils le trouveroient estrange. Quand j'ay fait quelques chose, je l'oublie presque entierement au bout de quelques mois, et plustost que de le chercher dans un chaos de brouillons que je n'ay pas le loisir de digerer, et de marquer par rubriques; je suis obligé de faire le travail tout de nouveau.' Ibid., ii, 227–32.

made numerous visits from the late 1670s until the middle of the following decade. His second great commission, which overshadowed his scholarly interests from the mid-1680s until his death, was to write a history of the Hanoverian dynasty; this commission led to a two-and-a-half year grand tour of southern Germany, Bohemia, Austria and Italy. His life from the 1690s onwards was characterised by frequent absence from Hanover and stays at the courts of Berlin and Vienna.

A further factor which contributed to the condition of his archive was the fact that Leibniz's scholarly work united official duties and private interests, with the two being intimately interwoven. This is a particular factor in relation to his correspondence, where no clear line of demarcation is possible between the correspondence of the civil servant and that of the scholar.

Of Leibniz's own administration of his papers little is known. There is no information on his having followed a particular archival system or method of storage of his papers. No general system of cover sheets, and no extensive lists or inventories by himself are known. There is likewise no indication of the state of his research papers based on accounts by visitors or colleagues during his lifetime.

In general biographers might be expected to throw light on the habits of the great person they study, particularly in relation to the management of their papers. To date, however, no definite biography of Leibniz, dealing comprehensively with all aspects of his life and scholarship and rooted in archival research, has been possible. No biographer has been able to claim to have examined all surviving Leibnitiana or even more than certain specific parts of Leibniz's archive. Moreover, no scholarly analytical investigation of the archive as a whole has been attempted. The remainder of this essay attempts to piece together what has been done to this end. Firstly, the development of the repository of Leibniz's archive, the former Royal Library, during his lifetime is examined. This is followed by an account of the fate of the repository, which included his manuscript papers and books, following his death. The final section outlines the efforts made since the eighteenth century to investigate the archive. Here interest focuses on the work of librarians and scholars in cataloguing the books and manuscript papers that has become the foundation for editorial projects over the last century and a half.

The development of the ducal library under the librarian and counsellor Leibniz

The history of the books and manuscript papers of Leibniz is closely connected with that of their present repository, the Niedersächsische Landesbibliothek (State Library of Lower Saxony) that succeeded the former Royal Library.[3] This library, over which Leibniz presided at his death, had developed from the ducal library of Johann Friedrich and had occupied several rooms of the official residence (the

3 W. Ohnsorge, *Zweihundert Jahre Geschichte der Königlichen Bibliothek zu Hannover (1665–1866)*, Veröffentlichungen der Niedersächsischen Archivverwaltung, no. 14 (Göttingen, 1962).

Leineschloss) when Leibniz first entered the service of the Duke in 1676. During his first years in Hanover Leibniz had occupied rooms adjacent to the Library. When Johann Friedrich died suddenly at the end of 1679 Leibniz was retained in the service of his brother and successor duke Ernst August. The new ruler however did not relish books and manuscripts as his brother had done; his main concern was not to promote scholarship but rather to increase the power and standing of his dynasty. Thus the library changed in character from being a scholarly encyclopaedic ducal library to an official 'bibliothèque de cabinet'.[4]

In the course of renovation of the castle at the beginning of the reign of Ernst August the ducal library was moved to a rear wing. This was the period of Leibniz's employment in the mining district of the Harz mountains and the library was effectively closed during the time of his engineering commitments (1679–85). When Leibniz was commissioned to write a history of the Hanover dynasty, the house of Guelphs, in 1685, the emphasis of the library changed and it became in effect the reference library of the historiographer. Between 1688 and 1690 the library was removed from the Leineschloss to a nearby private dwelling of the Anderten family in the Leinstrasse but remained closed until after Leibniz's grand tour of southern Germany, Austria and Italy (November 1687–June 1690).

In the 1690s the library was reserved exclusively for Leibniz's use. Ernst August (Elector of Hanover from 1692) desired that it be developed as a reference library for the history of the Guelphs and Leibniz enriched it by the acquisition of historical books and manuscripts. In 1698, at the beginning of the reign of Georg Ludwig as Elector of Hanover (from 1714 King George I of Great Britain and Ireland), the library and the librarian's apartment and rooms were removed to the house of the patrician Johann H. von Lüden in the Schmiedestrasse, the house that was to be Leibniz's residence until his death and was to become known to later generations of inhabitants of Hanover as 'Leibniz House'. The house survived until its destruction in bombing during World War II. Of the three princes Leibniz served under in Hanover, Georg Ludwig showed the least trust and confidence in him and least supported his scholarly activities.

At the time of Leibniz's death in 1716, the library personnel consisted of his collaborator, the historian Johann Georg Eckhart, an engraver, a bookbinder and two scribes all employed on the history project. Under Leibniz's directorship the library had been given an exclusive special purpose: a thematic emphasis in relation to the projected history of the Guelphs. This emphasis continued after his death and his successors in the eighteenth and early nineteenth centuries were to be both librarians and historiographers of the dynasty.

4 G. Scheel, 'Von der herzoglichen Bibliothek im Leineschloss zur Niedersächsischen Landesbibliothek an der Waterloostrasse. Eine Geschichte ihrer Standorte', in W. Totok and K.-H. Weimann (eds.), *Die Niedersächsische Landesbibliothek in Hannover. Entwicklung und Aufgaben* (Frankfurt am Main, 1976), pp. 70–90.

Leibniz, Eckhart and the opus historicum

Leibniz's collaboration with Eckhart was a significant influence on the extent and nature of the books and manuscripts he had amassed at the end of his life. Eckhart had been hired by Leibniz in 1698 as an assistant for his work on the history of the house of Guelphs.[5] There soon developed a collaboration, and indeed rivalry, that was to continue until Leibniz's death. In 1706 Leibniz helped Eckhart become Professor of History at the Hanoverian university in Helmstedt where he presented a doctoral dissertation in the same year. In 1711, however, Eckhart was released from teaching duties and recalled to Hanover to devote himself to the completion of the long overdue *opus historicum*. In effect, and much to the dislike of his superiors, Leibniz had departed from his original assignment to write a dynastic history and had expanded the work to be a history of the western Emperors of the Holy Roman Empire from 768 AD onwards. In 1714 Eckhart was appointed Hanoverian historiographer royal and, in 1715, Librarian under Leibniz's supervision. In Leibniz's service he undertook a great deal of spade work in terms of the collection of material for the history of the house of Guelphs and in the printing of the three-volume source collection published by Leibniz, the *Scriptores rerum Brunsvicensium* (1707–11).

Eckhart succeeded Leibniz as Hanoverian librarian and historiographer. In order to secure this appointment, Eckhart was not always loyal to Leibniz and defamed him, complaining of the slow progress of the *opus historicum* in letters to his superiors during the last two years of Leibniz's life. However, he showed merit especially in his efforts to secure posthumous publication of Leibniz's etymological and historical source material. Considerable debts and disagreements with the Hanoverian government, that refused any further subsidy for the publication of Leibniz's unfinished *Annales Imperii Occidentis Brunsvicensis* and even confiscated a biography of Leibniz he had written (published in 1779), induced Eckhart to flee from Hanover at the end of 1723 and enter the service of the Catholic Bishop of Würzburg the following year. Thus his efforts for the publication of the manuscript papers of Leibniz were terminated.

The fate of Leibniz's books and manuscript papers from the eighteenth to the twentieth centuries

Leibniz died at his apartment in the house of Frau von Lüden at the corner of the Schmiedestrasse and Kaiserstrasse, Hanover, on Saturday 14 November 1716 (Gregorian calendar) at about ten o'clock in the evening. At his deathbed were a servant and his last amanuensis Johann Hermann Vogler, who within minutes

[5] G. van den Heuvel, 'Johann Georg von Eckharts Entwurf einer Geschichte des Bistums Osnabrück', *Osnabrücker Mitteilungen. Mitteilungen des Vereins für Geschichte und Landeskunde von Osnabrück*, 101 (1996), 65–81, on pp. 65–8.

informed Leibniz's collaborator Eckhart. That same evening Eckhart, accompanied by a notary, closed off the apartment. The following morning the rooms of the deceased were officially sealed. Only after the arrival from Leipzig of Leibniz's nephew and heir, Simon Friedrich Löffler, could a royal commission appointed to examine the estate commence its work; this occurred on 26 November. Leibniz's remains were interred without pomp in a city church (the Neustädter Hof- und Stadtkirche St Johannes, as it is still called) on 14 December 1716.

The architecturally-imposing house in the Schmiedestrasse had accommodated Leibniz's apartment and his private book collection on the first floor, the Royal Library on the second floor and, on the top floor, Leibniz's study and working rooms. Among his personal belongings were found money and securities having a value of about 12,000 taler. An inventory of his manuscripts and correspondence, consisting of about 80 pages with about 500 headings, was prepared; this document is now preserved at the Hauptstaatsarchiv, the Main State Archive that serves as the public records office of the State of Lower Saxony in Hanover (Catalogue N°/Bench mark 93,6 No. 5, Vol. 7).[6] It was signed by Eckhart and the Secretary Türninger and represents the most authentic source for the condition of Leibniz's books and manuscript papers at the time of his death.

Leibniz's private collection of books in Hanover and Wölfenbüttel (where he had also served as librarian from 1691) numbered over 6,000, with a further 3,000 being disputed as his private property. The extent of the collection, which he never found time himself to catalogue, is documented in two inventories made in the course of the examination of his estate immediately after his death. These are preserved at the State Library of Lower Saxony in a fascicle of files with bench mark or catalogue number: 'Alte Bibliotheksakten A8'.[7] The library, according to the inventory, was arranged in a series of categories designated by the letters A to O. Although the categories were not entirely uniform, the following scheme represents the main division of the collection:

A. Jurisprudence	B. Medicine and science
C. Literature	D. Philosophy
E. Applied mathematics	F. Mathematics
G. Linguistics	H. Theology
J. Periodicals	K. Bibliographies
L, N, O Miscellaneous	M. History.

Leibniz had regarded his book collection as a tool or aid in his study and research. Evidence of his intensive use of these books is found in the form of a large number

6 G. Scheel, 'Leibniz als Historiker des Welfenhauses', in W. Totok and C. Haase (eds.), *Leibniz. Sein Leben – Sein Wirken – Seine Welt* (Hanover, 1966), pp. 227–76, on pp. 227–31.

7 A. Heinekamp, 'Leibniz' Privatbibliothek in der Niedersächsischen Landesbibliothek Hannover. Mit einem Titelverzeichnis der Abteilungen A (Jura) und D (Philosophia practica)', unpublished thesis (Cologne: Bibliothekar-Lehrinstitut des Landes Nordrhein-Westfalen, 1968).

of marginalia, underlinings and excerpts from books found among his manuscript papers.

Leibniz died unmarried and intestate and for a greater part of the eighteenth century a legal dispute raged between the Government in Hanover and the heirs to his estate.[8] Simon Friedrich Löffler, a clergyman of Leipzig, the son of Leibniz's sister Anna Katharina and her husband Simon Löffler, was recognised by the chancellery of justice as sole heir-at-law to the estate.[9] The protracted dispute between Löffler, his children and grandchildren, as heirs to Leibniz's estate, on the one hand, and the Government of Hanover, on the other, is documented among the Library records under the shelf-mark 'A8'.

In a resolution of 26 January 1717 George I decreed that the full estate of Leibniz should be handed over to his heir with the exception of Leibniz's private book collection, his manuscripts, correspondence and the four-function calculator on which he had worked over several decades. At the centre of the dispute was his private collection of books and his manuscripts, or, as far as the heirs to the estate were concerned, their equivalent pecuniary value. In the course of the dispute a catalogue of Leibniz's private book collection was prepared by Daniel Eberhard Baring, Eckhart and another librarian. This document, rediscovered in 1960, is now preserved at the State Library of Lower Saxony (Daniel Eberhard Baring, *Historische Nachricht von der Königlichen und Churfürstlichen öffentlichen Bibliothek in Hannover*, 1725; MS XXIII, N° 706a).[10] Successive petitions to the Government by Löffler in 1718, 1719, 1723 and 1725, which included demands for payment of a sum of money equivalent to the value of the books and manuscripts, were rejected. Following Löffler's death in 1748 his son Simon Friedrich Löffler junior (d. 1769) renewed the claims of his father demanding payment of a sum of 4,000 taler. As Leibniz's fame increased in the course of the century, concessions were drawn from the privy council. In 1768/9 a willingness to pay 500 taler was conceded, but it was the third generation of claimants who were to benefit from the settlement. Thus the matter was finally laid to rest in December 1779, sixty-three years after Leibniz's death, with the payment of 500 taler to four grandchildren of the original claimant.

Three years after Leibniz's death, the Royal Library, together with his private books, manuscript papers and calculator, was moved from the house where he died to the upper floor of the Archive building that was built between 1712 and 1719 and still serves as a public records office. The Archive and the Royal Library were to remain together for 170 years. Following renovation and extension of the building (completed in 1893) the Library was accommodated in a separate wing. At this time some of Leibniz's manuscript papers that had been separated from the

8 H. Lackmann, 'Der Erbschaftsstreit um Leibniz' Privatbibliothek', *Studia Leibnitiana*, 1 (1969), 126–36.

9 P. Schrecker, *G.-W. Leibniz. Lettres et fragments inédits sur les problèmes philosophiques, théologiques, politiques de la réconciliation des doctrines protestantes (1669–1704)* (Paris, 1934), pp. 16–19.

10 Ohnsorge, *Zweihundert Jahre Geschichte* (n. 3).

rest were retained in Hauptstaatsarchiv or Main State Archive, whereas the greater part were removed with the Library.

The main collection of Leibniz's manuscript papers survived political upheavals, war and occupations over a period of more than two and a half centuries relatively unscathed.[11] The Library, and the collection of Leibniz's books and manuscripts, suffered little or no loss during the years of French occupation (1803–14) in the Napoleonic era. The Library likewise survived the Second World War with relatively little loss or damage to its treasures, which had been removed for the most part to a depot outside Hanover city. Most of Leibniz's manuscript papers were preserved from 1939 until 1946 in the vault of a Hanover bank, the Landeskreditanstalt, where they survived the war intact. Severe flooding of 9–11 February 1946, that caused loss and damage to catalogues, preceded the return of the main body of manuscript papers at the end of April 1946. Today these manuscripts are preserved at the State Library of Lower Saxony in a building built between 1972 and 1975, at a distance of about half a kilometre from the Main State Archive and about one kilometre from the spot where Leibniz's house formerly stood.

Investigation of the manuscript papers of Leibniz since the eighteenth century

The Hanover archives today offer a unique record of both the work and the working of Leibniz.[12] Not only did he preserve all his own drafts, copies, extracts or abstracts of his letters, memoranda, notes, unfinished or rejected papers, etc.; he also saved all the letters he received from about 1,100 correspondents.[13] He did much the same with his administrative files, expense accounts, booksellers' and other bills, medical prescriptions, copies of historical documents and a mass of sundry papers. In short, he must have been reluctant to discard any sheet of paper with writing on it. Among his manuscript papers were writings on political matters and material assembled over nearly forty years on the history of the dynasty, in particular the *Origines Guelficae* and the *Annales Imperii Occidentis Brunsvicenses* which following repeated delays remained unpublished at the time of his death.

Notwithstanding multifarious official duties, Leibniz found time for extensive scholarly work in fields like philosophy, theology, mathematics, science, linguistics, jurisprudence and historiography. He corresponded with numerous persons of standing and fame, including princes, politicians and scholars. His universality in terms of interests, commented on by L'Hospital in 1693, and his extensive

11 O.H. May, *Kriegs- und Nachkriegsschicksale der Niedersächsischen Landesbibliothek in Hannover (1939–1950)* (Hildesheim, 1968).

12 K. Müller, 'Die Erschließung des Leibniz-Nachlasses', *Forschungen und Fortschritte*, 28 (1954), 57–60.

13 P. Schrecker, 'Qui me non nisi editis novit, non novit', *The Library Chronicle*, 20 (1954), 24–31.

network of correspondents are very much reflected in the extant manuscript papers. In all, the Leibniz archive comprises more than 60,000 items, including perhaps 16,000 letters (though this obviously depends on exactly what is defined as a letter); the whole comprises some 200,000 sheets of manuscript.

The collection can be analysed and seen to consist of different types or categories of writings as follows;[14] firstly, manuscripts of a great number of works, often accompanied by one or more drafts or copies made by his amanuenses, not seldom with marginal annotations or underlinings; secondly, excerpts from books, periodicals, documents and source material, in his own hand or those of his amanuenses; thirdly, a great number of notes and other writings in which he recorded his ideas for later investigations; fourthly, his correspondence, including the original letters and enclosures of his correspondents, often with endorsements by Leibniz himself, that vary in number from one to several hundred, and drafts or copies either by himself or his amanuenses of dispatched letters. Finally, in addition to the great number of marginalia in books, the surviving material includes his calculator, his travelling chair and a few smaller relics.

For a long time the very existence of the treasure of Leibniz's manuscript papers remained practically unknown to the learned world. When Louis Dutens edited the first *Leibnitii Opera omnia* in 1768, he included mainly previously printed writings, in fact everything that had appeared before Rud. Eric Raspe's publication of the *Nouveaux essais sur l'entendement humain* (1765). Only in the mid-nineteenth century was an effort made to introduce a systematic order to the mass of papers. Scholars and librarians, including Gottschalk Eduard Guhrauer, Johann Eduard Erdmann, Georg Heinrich Pertz, Alexander Louis de Foucher de Careil, Carl Immanuel Gerhardt, Onno Klopp, Louis Couturat and Eduard Bodemann, catalogued, ordered and edited parts of the unpublished treasure of manuscripts. The historian Pertz who was appointed Royal Librarian in September 1827 (and historiographer in February 1832) conceived a plan to publish posthumously not only the *Annales Imperii* of Leibniz but also all his unpublished historical and political works as well as his correspondence. This led to perhaps the first systematic attempt to arrange the manuscript papers of Leibniz beginning in 1837.[15] Political circumstances, however, led Pertz to quit Hanover and move to Berlin, from where he continued his edition, and to interruption of the arrangement of the Leibniz manuscripts in Hanover. In 1863, three years before the Prussian annexation of Hanover, Eduard Bodemann was appointed librarian. Under Prussian administration he was director of the library from 1867 until 1906 and his life's work became the cataloguing of its manuscript collections. His published catalogues, that are still the basis of the present-day arrangement of the main

[14] E. Bodemann, *Der Briefwechsel des Gottfried Wilhelm Leibniz in der Königlichen öffentlichen Bibliothek zu Hannover*, Reprographic reprint of first edition (Hanover, 1889), adds. and index G. Krönert and H. Lackmann, forw. K.-H. Weimann (Hildesheim, 1966), pp. v–x.

[15] Ohnsorge, *Zweihundert Jahre Geschichte* (n. 3), p. 64.

collection of Leibniz's manuscript papers in Hanover, provided a first calendrical description of more than 60,000 manuscripts consisting of letters[16]and writings.[17]

The collection is organised along the lines laid down by Bodemann. The main division is in two categories or groups designated LH ('Leibniz Handschriften') and LBr ('Leibniz Briefwechsel'). Within the first group, there is a further subdivision into forty-one subgroups according to the Bodemann scheme. Bodemann took account of the original arrangement as well as previous rearrangement of the material and so many Leibniz manuscripts in his catalogue retain their original catalogue number from the general manuscript catalogue of the library. In the second main division, where the letters are listed in alphabetical order according to the name of the correspondent, Bodemann also took account of earlier arrangement of the material and desisted from any further rearrangement. The Bodemann catalogues were republished in 1966 with additions, corrections and a name index for both volumes appended. In Bodemann's catalogue of 1889, 1,063 correspondents were included with more than 15,300 letters being taken into account; the second edition added a baker's dozen of additional correspondents, and the true number of correspondents may be in excess of 1,100.

The forty-one divisions of the Bodemann catalogue of writings (LH), are as follows:

I.	Theology	II.	Jurisprudence
III.	Medicine	IV.	Philosophy
V.	Philology	VI.	Geography
VII.	Chronology	VIII.	Genealogy/Heraldry
IX.	Archeology	X.	Numismatics
XI.	General history		

XII–XXXII.	Germany, Austria, German states, European and Asian states
XXXIII.	International law
XXXIV.	Politics and national economics
XXXV.	Mathematics
XXXVI.	Military matters
XXXVII.	Physics, mechanics, chemistry and natural history
XXXVIII.	Technology
XXXIX.	History of literature
XL.	Scientific societies, archives and libraries
XLI.	Leibniz's vita.

16 E. Bodemann, *Der Briefwechsel des Gottfried Wilhelm Leibniz* (n. 14).

17 E. Bodemann, *Die Leibniz-Handschriften der Königlichen öffentlichen Bibliothek zu Hannover*, reprographic reprint of first edition (Hanover and Leipzig, 1895), adds. and index G. Krönert and H. Lackmann, forw. K.-H. Weimann (Hildesheim, 1966).

Following Bodemann's scheme a forty-second division (XLII) has been added in this century. This includes manuscripts relating to the calculator, together with other material found or acquired in this century.

In effect none of the eighteenth- and nineteenth-century attempts to produce a complete edition of Leibniz's writings and correspondence was successful. Nonetheless a number of valuable editions did appear. The most important, which included the editions by Dutens of his *Opera Omnia* (1768), by Gerhardt of his *Mathematische Schriften* and *Philosophische Schriften* (1849–63 and 1875–90), by Foucher de Careil of his *Oeuvres* (1859–75) and by Klopp of his *Werke* (1864–84),[18] were presented by Emile Ravier in an appendix to his much acclaimed *Bibliographie des oeuvres de Leibniz* (1937) as 'Les grands éditions'.[19]

By the end of the nineteenth century it was recognised that an adequate edition would require the co-operative efforts of many specialists and the institutional facilities of an Academy or preferably of an international body of learned societies. In 1901 the International Association of Academies of Sciences took up the idea and entrusted the Academies of Berlin (the Royal Prussian Academy of Sciences) and Paris (the Académie des Sciences and the Académie des Sciences Morales et Politiques) with the execution of the plan to publish a complete and critical edition of all the extant writings of Leibniz. The first step, the creation of inventories of manuscript collections at Hanover and elsewhere as well as the necessary critical apparatus, got under way. Alas only two fascicles of the planned critical catalogue, covering Leibniz's early life to November 1676, were published under the direction of Paul Ritter[20] and Albert Rivaud.[21] The ruins of Franco-German relations that followed the Great War brought international co-operation to an end. The Berlin Academy adopted the project for a complete edition of Leibniz's letters and papers and it has continued under the aegis of this Academy, or its successors, to the present day. In connection with this undertaking the Berlin Catalogue (the so-called 'Berliner Arbeitsapparat') has been developed since the 1920s as a complete catalogue of the surviving letters and papers of Leibniz, both in Hanover and in other repositories throughout the world, from December 1676. This catalogue is continually updated and revised as new material comes to light. It forms the basis of all the editorial work on the Academy Edition of Leibniz's writings and letters now being conducted at three main centres in Germany (Hanover,

18 These editions are as follows: *G.G. Leibniz Opera Omnia*, ed. L. Dutens (6 vols., Geneva, 1768). G.W. Leibniz, *Mathematische Schriften*, ed. C.I. Gerhardt (7 vols., Berlin, later Halle, 1849–63; reprint Hildesheim, 1962). *Oeuvres de Leibniz*, ed. L.A. Foucher de Careil (7 vols., Paris, 1859–75; reprint Hildesheim, 1969). *Die Philosophischen Schriften von Leibniz*, ed. C.I. Gerhardt (7 vols., Berlin 1875–90; reprint Hildesheim, 1960–61). *Die Werke von Leibniz*, ed. O. Klopp, 1st series (11 vols., Hanover, 1864–84; vols. 7–11 reprint Hildesheim, 1970–73). *Leibniz: Sämtliche Schriften und Briefe*, ed. Preußische (later Deutsche) Akademie der Wissenschaften zu Berlin, 7 series, (29 vols. until end 1997, Darmstadt, later Leipzig, later Berlin, 1923–).

19 E. Ravier, *Bibliographie des oeuvres de Leibniz* (Paris, 1937; reprint Hildesheim, 1966).

20 *Kritischer Katalog der Leibniz-Handschriften*, ed. P. Ritter et al., no. 1 = 1. Heft (Berlin, 1908).

21 *Catalogue critique des manuscrits de Leibniz*, ed. A. Rivaud, no. 2 = fasc. II: March 1672 – November 1676 (Poitiers, 1914–24).

Münster, Potsdam).[22] At the Leibniz-Archiv, a department of the State Library of Lower Saxony, a group of scholars is editing the general, political and historical correspondence, the mathematical, scientific and technical correspondence and the mathematical writings of Leibniz.

The main body of Leibniz's manuscript papers, preserved at the State Library of Lower Saxony in Hanover, is also available on microfilm. A special library at the same location has been created to facilitate research on Leibniz's work and thought. The Niedersächsische Landesbibliothek today also houses the editorial office of the journal *Studia Leibnitiana*[23] that publishes an annual bibliography of literature on Leibniz. A cumulative bibliography in two volumes (to the year 1990) has also been published.[24]

Access to the main parts of the manuscript papers through microfilm copies is possible at at least two locations outside Germany. In the early 1950s, mainly through the initiative of Paul Schrecker, a project of microfilming the Hanover collection of manuscripts was initiated. The microfilm reels, reproducing about 100,000 pages, were deposited at the University of Pennsylvania Library.[25] These microfilms are now available to scholars at the Van Pelt Library of the University of Pennsylvania. A concordance compiled in 1955, linking each microfilm item to the Bodemann volumes, has recently been published in summary form.[26] A further collection of thirty-five reels is available at the University of London Library.[27]

Finally, at the end of the twentieth century, information science, supported by computer science and technology, seems to offer new opportunities of accessing and investigating large collections of manuscript papers like those of Leibniz. It remains to be seen, however, whether, and in what time scale the new technologies will substantially contribute to increased accessibility of such manuscript collections for scholars and editors.

22 Leibniz, *Sämtliche Schriften und Briefe* (n. 18).

23 *Studia Leibnitiana. Vierteljahrschrift für Philosophie und Geschichte der Wissenschaften*, publ. Gottfried-Wilhelm-Leibniz Gesellschaft e.V. (Wiesbaden, 1969–).

24 *Leibniz-Bibliographie. Die Literatur über Leibniz bis 1980. Begründet von Kurt Müller*, ed. A. Heinekamp, 2nd ed. (Frankfurt am Main, 1984). *Leibniz-Bibliographie. Band 2. Die Literatur über Leibniz 1981–1990*, ed. A. Heinekamp et al., vol. 2 (Frankfurt am Main, 1996).

25 Schrecker, 'Qui me non nisi editis novit' (n. 13).

26 P. Lodge, 'Leibniz microfilms at the University of Pennsylvania', *Leibniz Society Review*, 6 (1996), 164–9.

27 L.M. Newman, *Leibniz (1646–1716) and the German library scene*, Library Association Pamphlet, no. 28 (London, 1966), on pp. 7 and 38.

11

Of Records and Grandeur: the Archive of the Royal Society

Mordechai Feingold

> Their purpose is, in short, to make faithful *Records*, of all the Works of *Nature*, or *Art*, which can come within their reach: that so the present Age, and posterity, may be able to put a mark on the Errors, which have been strengthened by long prescription: to restore the Truths, that have lain neglected: to push on those, which are already known, to more various uses: and to make the way more passable, to what remains unreveal'd. This is the compass of their Design.

Thus in 1667 Thomas Sprat enunciated one of the professed aims of the Royal Society. Nor was he the only one to articulate such an ideal. Sir Robert Moray's 1663 memorandum on 'The business & designe of the Royal Society' specified that its first aim was 'To improve the knowledge of Naturall things and all usefull Arts . . . by Experiment', its second 'To attempt the recovering of such allowable arts & inventions as are lost'.[1] In 1668 Robert Hooke elaborated further on the significance of the Society's latter enterprise of collecting records:

> I am not, I suppose, the only Person who hath heard some persons . . . ask what the *Royal Society hath done for so many Years* as they have met . . . it might be a satisfactory Answer to such Objectours that this *Society* have been imployed in collecting such *Observations*, and making such *Experiments* and *Trials* as being fitly apply'd and judiciously made use of, will very much tend to the advancement of *Natural Knowledge*: And tho' the things so collected may of themselves seem but like a rude heap of unpolish'd and unshap'd Materials, yet for the most part they are so qualified as that they may be fit for the beginning, at least of a solid, firm and lasting Structure of Philosophy. But because some of those may doubt whether really there be any such Collection made . . . I conceive, it may not be altogether unseasonable this following Year nor improper for this *Honourable Society's Care* to make some attempt of that kind by shewing some *Specimen* of such a Structure

I would like to thank Michael Hunter, Mary Sampson, and David Sturdy for their helpful comments.

1 Thomas Sprat, *History of the Royal Society* (1667), eds. Jackson I. Cope and Harold W. Jones (London, 1959), pp. 61–2; Michael Hunter, *Science and the Shape of Orthodoxy* (Woodbridge, 1995), p. 172.

> raised from Observations and Collections of their own, that it may appear that they have not disquieted themselves in vain, in heaping up such a Treasure which they know not who shall enjoy or make use of; that is, to see whether any of these things they have been collecting, will afford sufficient Evidence to ground a deduction of a higher Nature upon.[2]

Such statements make it clear that in adopting a Baconian 'programme' as its preferred mode of operation, the Royal Society understood the concomitant requirement to establish and maintain an archive that, in Bacon's vision, would house the huge collections (both literal and material) of particulars, 'as may serve for a foundation to build philosophy upon'. Moreover, in implementing this Baconian agenda the Society embarked on a course already advocated by several earlier schemes. Thus, for example, Francis Kynaston's *Musaeum Minervae* had contemplated serving as a small-scale research institute and repository of scientific information. 'The Professour of *Philosophie* and *Physick* and his Assistants', wrote Kynaston,

> shall from time to time according to their abilities and opportunities make experiments of naturall things, chiefly of medicinall use, and what they finde certain shall be recorded, and what they finde otherwise shall in a book by it self be noted, how and in what manner experiment was made, and how it failed, that men may not afterwards spend their pretious time and meanes in vain, and that the licentious abuses of *Impostures* hereafter may be detected.

For his part, 'the *Astronomicall* Professour shall keep an exact *Diary* of every of the *Coelestiall Appearances* and of the weather, that at last we may finde the causes of our insularie varieties. And every moneths observations shall be fairely written out, and given up to the Library.'[3] Less detailed arrangements for the keeping of records as well as the Baconian collection can be found in the proposals of both Abraham Cowley and John Evelyn, the latter of whom also contemplated a spacious room in his draught design of the college.[4]

To the intention of the Society to collect experiments and observations can be added its commitment to an archive that would contain at the very least summaries of all that had been performed and previously discovered. Already in the early 1650s the members of the Oxford Philosophical Club had begun such a project. William Petty wrote to Hartlib on 12 February 1651/2 to inform him that

> The Club-men have cantonized or are cantonizing their whole Academy to taske men to several imploiments and amongst others to make Medullas of

2 *The Posthumous Works of Robert Hooke*, ed. Richard Waller (London, 1705, repr. London, 1971), p. 329.

3 Francis Kynaston, *The Constitutions of the Musaeum Minervae* (London, 1636), pp. 7, 12.

4 For Evelyn's proposal see Thomas Birch (ed.), *The Works of of the Honourable Robert Boyle*, 2nd ed. (6 vols., London, 1772, repr. Hildesheim, 1966), vi, 288–91. The drawing is reproduced in Michael Hunter, *Establishing the New Science: The Experience of the Early Royal Society* (Woodbridge, 1989), pp. 182–3.

all Authors in reference to experimental learning. Thus they intend to doe with Kircherus Workes and others whatsoever.

Seth Ward wrote more fully two weeks later, on 27 February 1652:

> we have (every one takeing their portion) gone over all or most of the heads of naturall philosophy & mixt mathematics collecting onely an history of the phenomena out of such authors as we had occasion and opportunity, our first business is to gather together such things as are already discovered and to make a booke with a generall index of them, then to have a collection of those which are still inquirenda and according to our opportunityes to make inquisitive experiments, the end is that out of a sufficient number of sure experiments, the way of nature in workeing may be discovered, but because (not knoweing what others have done before us) we may probably spend our labour upon that which is already done, we have conceived it requisite to examine all the bookes of our public library (everyone takeing his part) and to make a catalogue or index of the matters and that very particularly in philosophy physic mathematics & indeed in all other facultyes, that so that greate numbers of books may be serviceable and a man may at once see where he may find whatever is there concerneing the argument he is upon, and this is our present business which we hope to dispatch this Lent.[5]

And just as the Society adopted various other procedures advocated by earlier schemers, so, too, it made an effort to scrutinise printed books for scientific and useful information. As Sprat put it

> Having well succeeded in this their purpose of *collecting* divers patterns of all *Natural*, and *Artificial* things: they have also (amongst others) appointed a *Committee*, whose chief employment shall be to read over whatever *Books* have been written on such subjects. By this means they hope speedily to observe, and digest into *Manuscript volumes*, all that has been hitherto try'd, or propounded in such studies. This is the only help that an *Experimenter* can receive from *Books*: which he may still use, as his *Guide*, though not as his *Masters*. For this end they have begun a *Library* consisting only of such *Authors*, as may be serviceable to their *Design*.[6]

Thus, the concept of the archive was integral to the very idea of institutionalising a scientific society fashioned, as was the Royal Society, along the Baconian model. And although the concept of an archive did not originate with the Royal Society, its members were nonetheless the first to attempt to implement this ideal on a large scale. That they were not entirely cognisant of the enormity of their enterprise is evident from the early minutes. At first they apparently believed they could appoint one of the Fellows, on a rotating basis, as 'register' who, assisted

5 H.W. Robinson, 'An Unpublished Letter of Dr. Seth Ward Relating to the Early Meetings of the Oxford Philosophical Society', *Notes and Records of the Royal Society*, 7 (1949), 69–70; G.H. Turnbull, 'Samuel Hartlib's Influence on the Early History of the Royal Society', *Notes and Records of the Royal Society*, 10 (1953), 101–30, on p. 113.

6 Sprat, *History* (n. 1), p. 252.

by an amanuensis, would 'provide three books, one for the statutes of the society and names of its members; another for experiments, and the result of debates; and a third for occasional orders'. In addition, 'three or more of the society be desired to be reporters for that meeting, to sit at the table with the register, and take notes of every thing of importance, that shall be offered to the society, and debated in it; and that they together form a report against the next meeting, to be filed by the register'.[7] Before long, however, the members realised that a far more professional attitude was required. As Sprat described the transformation

> They had at first a *Register*, who was to take Notes of all that pass'd; which were afterwards to be reduc'd into their *Journals*, and *Register Books*. This Task was first *perform'd* by Dr. *Croone*. But they since thought it more necessary, to appoint two *Secretaries*, who are to reply to all Addresses from abroad, and at home; and to publish whatever shall be agreed upon by the *Society*.

In 1668 Oldenburg himself commented at length on what was required from a diligent secretary:

> He attends constantly the Meetings both of the Society and Councill; noteth the Observables, said and done there; digesteth them in private; takes care to have them entered in the Journal and Register books; reads over and corrects all entrys; sollicites the performances of taskes recommended and undertaken; writes all Letters abroad and answers the returns made to them, entertaining a correspondence with at least 30. persons; employes a great deal of time, and takes much pains in inquiring after and satisfying forrain demands about philosophicall matters, dispenseth farr and near store of directions and inquiries for the Society's purpose, and sees them well recommended etc. Query. Whether such a person ought to be left un-assisted?[8]

Oldenburg, in fact, proved himself virtually irreplaceable. However, the difficulty of coping with an ever-expanding archive could not be resolved simply by finding qualified professionals. When analysing the nature and scope of the archive, one should remember that for the first fifty years of its existence the Society lacked a permanent home, thus exacerbating the already substantial problem of its keeping records. According to Sprat, the Gresham College authorities allowed the Society 'one *publick Room* to meet in, another for a *Repository* to keep their Instruments, Books, Rarities, Papers, and whatever else belongs to them: making use besides, by permission, of several of the other Lodgings, as their occasions do require'.[9] Whatever the exact nature of the arrangements, they were

7 Thomas Birch, *The History of the Royal Society of London* (4 vols., London, 1756–57; repr. Bruxelles, 1967), i, 7. The amanuensis, too, was now expected 'to attend every meeting-day, and set down such things as the society shall think proper; and that his annual salary be increased from two to four pounds': ibid., i, 17.

8 Sprat, *History* (n. 1), p. 94; Oldenburg, *Correspondence*, iv, xxiv. Oldenburg prepared this document to justify his supplication to the Society to remunerate him properly.

9 Sprat, *History* (n. 1), p. 93.

less than satisfactory for the secretaries' use. Likewise, in contrast to the librarians (first Goddard and then Hooke, who took charge of the repository as well) who, as Gresham professors, benefitted from spacious quarters in the College, the secretaries often had to take home with them many of the Society's books and papers in order to discharge the time-consuming tasks of entering records, copying minutes and carrying on correspondence. The general state of flux of the Society's papers is illustrated by Henry Oldenburg who, faced with the possibility of contracting the plague, ruminated on the fate of the papers in his possession. 'In the meane time,' he wrote Boyle in early July 1665

> I am not a little perplexed concerning the Bookes and Papers belonging to the Society, that are all in my Custody. All I can think of to doe in this case, is, to make a list of them all, and to putt them up by themselves in a boxe, and seale them together with a superscription; that so, in case the Lord should visit me, as soon I find myselfe not well, it may be ready to be immediately sent away out of mine to a sound house.

Six weeks later he finished the task and deposited them in a separate box, which he proceeded to lock.[10]

Under such circumstances the retrieval of records belonging to the Society could prove problematic, especially if an officer died while in office. Such was the case following Oldenburg's death in 1677. Though Hooke and other fellows managed to reclaim most of the Society's papers from Oldenburg's widow, some appear to have been missed, for at a meeting of Council on 3 January 1683 it was noted that 'The Earl of Anglesey having formerly bought Mr. Oldenburg's books and papers, and it being supposed, that his Lordship had received several of the Society's papers therewith, it was ordered, that his Lordship should be spoken to, either to restore them, or let copies be taken.' The outcome of this matter is obscure; however, it prompted a concerted effort to implement tighter procedures for safeguarding the Society's papers:

> The secretaries having drawn up a state of the papers in their office, and it appearing, that several things of the Society's had not been put into their hands, as ought to have been; it was ordered, That Mr. Hooke deliver up into the hands of either of the secretaries all such books and papers, as any way belong to the Society, or came to his hands upon the account of his having been secretary.[11]

Nevertheless, as the secretaries still lacked a proper office – and notwithstanding the decision of 24 September 1677 that 'all papers and books concerning the Society be kept in the Repository or Library of the said Society; and that if any thing be to be transcribed, it be done there' – their working habits remained unchanged. In February 1686, when one of the Journal books 'was wanting in the

[10] Oldenburg, *Correspondence*, ii, 430, 459, 592.
[11] Birch, *History* (n. 7), iii, 342; iv, 174.

Society's presses, Mr. Aston said, that he had it in his possession; as likewise all the copies of the journals and registers except two'. Henry Hunt was ordered to retrieve them, apparently without success, for a month later 'It was ordered, that Mr. Halley should wait upon Mr. Aston about the copies of the journal and register books of the Society in his hands; as likewise to have from him the original papers and letters belonging to the Society.'[12]

Clearly, the increasing difficulty of handling an ever-expanding body of documents, compounded by the vicissitudes of the early years, had important repercussions on the content of the archive. Other factors that contributed to determining its present shape included a general lack of clarity among contemporaries over what constituted an official document and, more importantly, what documents ought to be preserved. Obviously the Journal Books, Council minutes and Register were unambiguously official records. But what about letters? Oldenburg certainly recognised the indispensability of his correspondence for the Society's mission and reputation, and the Halls' edition is a testimony to the manner in which he acted on this belief. Thus, on 5 December 1667, when Oldenburg initiated a discussion on

> how the letters, which belonged to the Society, might be well preserved, it was resolved, that the Council should be desired to take care, that all letters, which are written by and to the Society, or by and to any of their members, be put up together in a book, and thence transcribed into the Letter-book appointed by statute for that purpose.[13]

Likewise, Sir Robert Moray appreciated the need to preserve at least some of the letters he received, and to this end in 1668 he asked John Collins 'to peruse all such letters as Huygenius formerly wrote to him, and they are now in Mr Oldenburgh's hands, to be transcribed, and remaine upon Record in the Royall Societie'. Periodically, Council continued to order the preservation of letters sent to the Society, or even copies of letters sent by the secretaries – though invariably such actions came in the wake of some mishap. For example, when Abraham Hill asked at a meeting of the Society on 5 April 1682 that a letter sent previously to Leeuwenhoek be read, and Hooke

> not knowing whether he had any copy of it, Sir Joseph Williamson and Mr. Hill were very urgent to have all papers entered into books. The Society thereupon thinking it a matter not so proper to be discoursed of at these meetings, desired, that it should be recommended to the council to consider of it at their next meeting.

As happened so often, the matter was eventually dropped.[14]

Some of the more conscientious members went further than simply considering

12 Ibid., iii, 343; iv, 457, 462.
13 Ibid., ii, 223–4.
14 Herbert W. Turnbull (ed.), *James Gregory: Tercentenary Memorial Volume* (London, 1939), pp. 54–5; Birch, *History* (n. 7), ii, 223–4; iv, 141.

the preservation of registers and official papers submitted to the Society, envisaging the Society also as the proper repository of all scientists' papers. As early as 1662, for example, we encounter Sir Robert Moray attempting to use the good services of Matthew Wren in retrieving the papers of Thomas Harriot, then in the possession of Edward Hyde, Earl of Clarendon. Wren made some inquiries concerning the papers and informed the Society that Clarendon was indeed willing to communicate them, but apparently the matter was not pursued further. Seven years later John Collins mentioned that many Harriot papers 'were in the hands of the son of the earl of Cherbury'. Once again, notwithstanding Oldenburg's promise to try and 'procure a sight and transcript of them', nothing was accomplished.[15]

Collins, in particular, was indefatigable in attempting to conserve the papers of departed practitioners in order to ensure their priority claims as well as make their works accessible to future generations. Thus, in 1669 he urged the Society to pursue the papers of Samuel Foster, formerly Gresham Professor of Astronomy, but had little success.[16] In 1675 Collins was even more determined to ensure the preservation of the legacy of his good friend James Gregory. As he wrote to David Gregory, James' brother

> I have drawne up an accompt of the Letter Commerce and of all that was Mathematicall that I at any time received from your Brother, either in writing or in discourse, with an intent to impart the same to the Royall Society whereof he was a Member, where any friend of his may peruse the same or have it transcribed . . . as to the sincerity of doing whereof without omitting anything materiall that was not controversiall with Hugenius etc., I can safely touch the Altar.[17]

John Aubrey, too, was committed to such an ideal. Having obtained in 1684 the Society's encouragement, he managed to procure the papers of the mathematician Thomas Merry – which also included works by Briggs and Gellibrand. An inventory was made and the papers were designated for deposit in the archive, but apparently something went awry, and the papers never made it to their intended destination. One wonders whether the episode so discouraged Aubrey that he no longer saw the Society's archive as the preferred repository of his own papers. As he wrote shortly thereafter to William Musgrave, secretary of the Oxford Philosophical Society, he intended to deposit with them 'two or 3 considerable letters worthy the keeping of so learned and carefull a Societie: I dare not trust them with ours, we are so careles'.[18]

15 Ibid., i, 120, 126, 309; ii, 410. Moray also attempted in 1662 to procure some of William Harvey's papers, also said to be in Clarendon's hands. Nothing more is heard of this. Ibid., i, 120. The reference to the Earl of Cherbury, adds Birch, 'is a mistake probably for Richard lord Vaughan and earl of Carbery; or for the lord Herbert of Cherbury'.

16 Seven years later John Aubrey was able to procure such papers for the Society's inspection, and a list of them was read at one of the meetings, but this is the last we hear of them. Ibid., ii, 410; iii, 316.

17 *Gregory Tercentenary Volume* (n. 14), p. 344.

18 Birch, *History* (n. 7), iv, 223, 224, 296, 326–7; R.T. Gunther (ed.), *Early Science in Oxford* (14 vols., Oxford, 1921–45), xii, 269; Michael Hunter, *John Aubrey and the Realm of Learning* (London, 1975), pp. 64–5.

Given this rather befuddled situation, no wonder that even the secretaries' correspondence often survives in archives other than those of the Society, including many public and private libraries throughout England. The Newton archives at Cambridge and the Sloane collection in the British Library, for example, contain not a few 'official' Society records, in addition to many letters addressed to Newton and Sloane in their official capacity, which they considered as private property. The surviving papers of many Fellows also include material that for some reason is not duplicated in the Society's archive, such as letters from English and foreign correspondents who transmitted information for the Society's use. Such letters were often read at a Society meeting but not always preserved. In addition, we find scattered in various archives reports on books, experiments, and accounts that the Society commissioned from members and for which no copies exist today in the Society's archive despite the fact that at least some were read during one of the weekly meetings.

In addition to the disappearance or dislocation of records that had once belonged to the archive, we are faced with the problem that some records have survived, but not in their original form. A close study of the council minutes demonstrates that sections of the Journal Books and Registers were recopied or emended on more than one occasion, and in the process their content may have been edited. To give but one example. On 28 March 1683

> The council having taken notice, that the journal books of the meetings of the Society, which have been written since 1677, have several vacancies in them, omissions of things and names, and mistakes; it was ordered, That any three or more of the council, then present, or of any other members of it for the present year, be a committee to meet at the repository, and inspect the journal books, to note any vacancies, omissions, or mistakes; or, Mr. Hooke being present, to meet them, and supply them, with his consent or opinion, and on just occasions alter and strike out such places or passages, as he and they shall agree to; and in case of disagreement, make report to the council, that they may take farther order therein; and if the omissions and vacancies be such, as could not be supplied, then to draw lines there in void spaces, that for the future there may be no new thing written therein.

Four months later we read: 'The journal books being not yet corrected and perfected, according to an order of March 28, the president was pleased to undertake the trouble of doing it; and the journal books were left at the repository for his use.'[19] Though nothing more is heard of this matter, other references to the copying of the various books, and of payments to those engaged in the task, should caution us against assuming that these records always retained their original form.

Indications of wilful 'tampering' also exist. At the height of the dispute between Hooke and Huygens, Oldenburg must have felt somewhat anxious – if not downright guilty – for on two occasions at least he obliterated several lines from Francis Vernon's letters to him that appeared to suggest that there existed a

[19] Birch, *History* (n. 7), iv, 196, 213.

particularly close tie between him and Huygens which, in turn, could be interpreted as proof of his championing the latter's priority claims.[20] Hooke was also involved in an attempt to alter records, this time in 1684 when John Flamsteed demanded 'that the journal-book in 1682 might be altered as to some expressions reflecting upon him entered by Mr. Hooke. It was referred to another time.' A few months later, following another complaint by Flamsteed

> that he had been reflected upon by Mr. Hooke in the minutes of the Society, it was ordered, that a line should be drawn through the places complained of, and that there should be written on the side, *cancelled by order of council*: and that the journal-book should be brought to the next meeting of the council, who should see it done.[21]

If the order was observed, a subsequent copying of the volume in question removed the evidence.

Charges of deliberate failure to register members' contributions represent a different sort of 'tampering' with the records. But Hooke was not alone to level such charges against Oldenburg. Already Thomas Sprat, who claimed in his *History* that he was 'cautious to forbear Commending the labours of any private *Fellows* of the *Society*', was forced to break his resolve in the case of Christopher Wren, not because of his intimate friendship with the latter but 'on the meer consideration of Justice: For in turning over the *Registers* of the *Society*, [he] perceiv'd that many excellent things, whose first *Invention* ought to be ascrib'd to him, were casually omitted'. Consequently Sprat was prompted 'to do him right by himself, and to give this separate Account of his indeavours, in promoting the Design of the *Royal Society*, in the small Time wherein he has had the opportunity of attending it'. Wren's grandson was even more explicit in his accusations:

> Mr. *Henry Oldenburg*, the first Secretary to the *Royal Society*, with Disingenuity, and Breach of Trust, communicated, and calandestinely convey'd into foreign Parts, particularly *Germany* and *France*, divers of the Inventions, and original Experiments of the *Author*; which were afterwards unfairly claim'd by others, as the true Inventors, and publish'd abroad under other Names.[22]

The realisation of the centrality of record-keeping for the scientific mission of the Society was accompanied by the propagandist need to advertise their accomplishment. The pride with which the Society exhibited its contribution to the accumulation and preservation of records is evident in another passage by Sprat:

> The *Society* has reduc'd its principal observations, into one *common-stock*; and laid them up in publique *Registers*, to be nakedly transmitted to the next Generation of Men; and so from them, to their Successors. And as their

20 Oldenburg, *Correspondence*, vi, 502, 504–5.

21 Birch, *History* (n. 7), iv, 277, 318.

22 Sprat, *History* (n. 1), p. 311; Christopher Wren, Jr, *Parentalia: or, Memoirs of the Family of the Wrens* (London, 1750), p. 199.

purpose was, to heap up a mixt Mass of *Experiments*, without digesting them into any perfect model: so to this end, they confin'd themselves to no order of subjects; and whatever they have recorded, they have done it, not as compleat Schemes of opinions, but as bare unfinish'd Histories.

'By their fair, and equal, and submissive way of *Registring* nothing, but *Histories*, and *Relations*,' Sprat continued, 'they have left room for others, that shall succeed, to *change*, to *augment*, to *approve*, to *contradict* them, at their discretion. By this, they have given *posterity* a far greater power of judging them; than ever they took over those, that went before them.'[23]

Notwithstanding Sprat's eloquent and self-congratulatory account, the professed openness of the archives and the purported design to make knowledge accessible to all potential researchers is contradicted by the official policy of the Society. In fact, Hobbes's complaint that 'the place where they meet is not public'[24] is perhaps more pertinent in this context for unlike the quick reversal of the decision to limit membership, the archive continued to remain the private domain of the Fellows. As early as March 1661 'It was resolved, that it should be free for any member of the society to have copies of any paper registered, but not to communicate them to strangers without leave of the society, except it were his own paper.' The policy was enshrined in the 1663 statutes of the Society:

No Fellow shall give any copy or transcript of any matter contained in the Register-books, or other Books of the Society, to any one that is not a Member thereof; nor communicate the same to any such person to be transcribed, without particular leave obtained at a Meeting of the Society; except the said matter were originally brought in, or communicated by himself.

For his part, the clerk was admonished not to allow anyone who was not a member 'to read any book, record, or other writing, or any part thereof, belonging to the Society; nor give any copies thereof, nor any way communicate any thing contained therein, to any such person: and in case he shall do contrary thereunto, he shall forfeit his place'.[25]

Once again, the Society had simply implemented procedures that had been formulated earlier. The constitutions of the Musaeum Minervae, for example, stipulated that 'the Bookes of *Secrets*, *Experiments*, and *Demonstrations* shall be communicated to none but to the Professours, *septennalls* and Receiver, or such other, as the Regent and Proffessours shall upon very great reason see cause to yeeld unto'. Similarly, Abraham Cowley's plan of a college specified that 'there shall be kept a Register under lock and key, and not to be seen but by the Professors, of all the Experiments that succeed, signed by the persons who made the trial'.[26]

23 Sprat, *History* (n. 1), pp. 115–16.

24 Thomas Hobbes, *Physical Dialogue*, trans. Simon Schaffer, in Steven Shapin and Simon Schaffer, *Leviathan and the Air-Pump* (Princeton, 1989), p. 350.

25 Birch, *History* (n. 7), i, 18; *The Record of the Royal Society of London* (4th ed., London, 1940), pp. 300, 298.

26 Kynaston, *Musaeum Minervae* (n. 3), p. 17; Abraham Cowley, *A Proposition for the Advancement of*

Despite what at first glance might appear to be a policy of secretiveness, the willingness of the Society to advertise and, later, publish detailed reports of accounts submitted and read at the meetings – not to mention later proposals to print indices to the Journal Books, Registers, and Letter Books – suggests a very different motive than the one at work, say, in the early Académie des Sciences. (Intentions aside, the size and constitution of the membership would have made it in any case impossible to impose confidentiality.) Limiting access to the archive was intended primarily to encourage potential practitioners to entrust their discoveries to the Society as well as to stake the monopolistic claims of a new and vulnerable corporate institution on a specialised form of knowledge. In contrast, Robert Hooke's proposals of reform during the 1670s embraced secrecy as the preferred mode for the Society's operation. He proposed that the Fellows would pledge, under oath, to keep to themselves much of what they learned at the meetings, further suggesting that the Society print for the sole use of non-London based Fellows accounts of what had transpired at the weekly meetings.[27]

Michael Hunter has found something of a discrepancy between the professed ambition of the Society to maintain and exploit an extensive archive and the actual use made of it by contemporaries – including its Fellows. My own reading is that such a conclusion may be too severe. Although much of the evidence concerning the use made of the archives by its members has long since vanished, much can be inferred, if not be established, from the surviving documents. Thus, in addition to the extensive copies of the minutes and registers sent by Oldenburg to Boyle, we have samples of exact transcripts of the Journal books sent to Martin Lister between February and June 1673, or transcripts of Sluse's and Huygens' letters sent to Pell.[28] In addition, it is obvious that Oldenburg made extensive transcripts of the minutes and registers for the use of such faithful correspondents as John Wallis and John Beale, though virtually all of Oldenburg's letters to them are lost.[29] Nor was Oldenburg alone in disseminating archival material. During the early 1660s Sir Robert Moray sent Christiaan Huygens extremely detailed accounts of

Learning (1661), in *The Complete Works in Verse and Prose of Abraham Cowley*, ed. Alexander B. Grosart (2 vols., Edinburgh, 1881, repr. Hildesheim, 1969), ii, 289.

27 Birch, *History* (n. 7), iv, 251; Hunter, *Establishing the New Science* (n. 4), pp. 199–201, 237–8: The discourse and experiments, Hooke wrote, shall be 'transcribed into the Register book there to be perused by any member of [238] the Society at any seasonable time & by none els whatever. The breviate of all the transactions of the Society performed at their usuall meeting shall be carefully made perused & corrected by a select number of the members chosen pro tempore by the society, and once a month or oftner if the matter requires there shall be ( ) printed in the forme of a gazet which by the Register shall be sent to every member of the S[ociety] which is absent provided he take care to pay for the postage & conveyance thereof. and if upon seeing the breviat he desireth to be further satisfyd by having the whole Discourse it self to peruse[,] the Register shal take care to have it transcribed provided he defray the charge thereof after the rate of 04^{d} for every thousand words'.

28 Hunter, *Establishing the New Science* (n. 4), pp. 37–8 n. 133; Royal Society, Boyle Papers 20; Bodleian MS Lister 34, fols. 51, 53–6, 89–91v, 93, 100, 106–106v. Other examples include RS MS 215 – copies of papers sent during the 1660s and 1670s to an anonymous member. See also BL MSS Sloane 243, 698.

29 Oldenburg, *Correspondence*, passim. See, for example, Oldenburg's letter of 3 December 1664 to Boyle, enclosing Huygens' 'account concerning an universall measure, which I beg may be communicated to Dr Wallis . . . and whereupon I expect both your sense'. Ibid., ii, 328.

the Society's meetings, almost on a weekly basis, a practice that became less detailed and more erratic in later years when Oldenburg assumed charge of the epistolary exchange with Huygens. For his part, John Collins also made both direct and indirect use of the Society's books. Writing to James Gregory on 23 February 1671/2, for example, Collins enclosed 'an accompt of Mr Newton's Tellescope', having 'obtained the very words of his Proposall out of the Journall of the R. Societie'.[30] Indeed, much of the correspondence of the secretaries should be viewed as making use of the archive, for the information conveyed often encapsulated material derived either from the weekly meetings or from information received from other correspondents.

Archival records were also used for purposes other than informing distant members of the proceedings in London. They were regularly searched in order to establish priority claims or to retrieve previous discussions of certain subjects. In 1668, for example, Collins informed Gregory that 'Huygens lately sent over his Theories of Motion, Percussion and Collision wherein it seemes nothing more is discovered then what Dr Wren formerly imparted, diverse yeares since to the Societie (as appears per their Records).' Similar searches were carried out in early 1686. When on 17 February Hooke claimed that 'Mr. Hubin's contrivance of the barometer was formerly brought in to the Society by himself, and that he in truth was the inventor thereof', Halley 'was ordered to search the books of the Society, to see, whether there be any record thereof'. Five weeks later he was able to report back that 'It appeared from the journal of the Society, that the contrivance of the barometer made and published by Mr. Hubin in the year 1673, was brought in to the Society by Mr. Hooke in June 1668.' (The event, incidentally, is not recorded in the Society's minutes.) The following month another search of the Register Book revealed 'that Mr. Hooke's way of carrying air under water for the use of divers by a chain of buckets inverted, brought in by him in the year 1663, was not entered there'.[31]

Hooke certainly made extensive use of the records. Following Oldenburg's death he scrutinised them carefully in order to prove the latter's connivance with Huygens; similarly, in the mid-1680s he was given permission to search both Journal-books and Council minutes when preparing his legal case against Sir John Cutler. A different sort of search was attempted by Tancred Robinson in 1708, on behalf of Martin Lister: 'I enquir'd of Dr Sloan, Mr Aston, Dr Thorp, and Mr Waller', Robinson wrote, 'if they could informe me of any particulars (not printed) in the Registers of the R. S. relating to the Exp. of weighing the blood of Animals, but upon examination they could tell me nothing distinct from what Dr Moulins, yourself and Dr Keil have already publish'd.'[32]

Last but not least, the archive could be made use of for more 'sinister' purposes.

30 Huygens, *Oeuvres complètes*, vols. iii–vi, passim; *Gregory Tercentenary Volume* (n. 14), p. 219.

31 Ibid., p. 65; Birch *History* (n. 7), iv, 459–61, 464.

32 H.W. Robinson and Walter Adams (eds.), *The Diary of Robert Hooke, 1672–1680* (London, 1935, repr. London, 1968), pp. 324, 338; Hunter, *Establishing the New Science* (n. 4), pp. 316–30; Birch, *History* (n. 7) iv, 210, 229; Bodleian MS Lister 37, fol. 132.

One of Hobbes's grievances in the *Dialogus physicus* of 1661 was that the Fellows of the Society conspired to disgrace him both now and to all posterity by deliberately preserving in their archive results that he had disavowed: 'I thought to have found some method of interposing two mean proportional lines between any two straight lines', Hobbes wrote, but having sent it to a friend he discovered he was wrong and he 'wrote a recantation of it'. John Wallis, 'seeing the same fault in the meantime, which was easily done, refuted it. They reproduced this refutation in the archives of the Society, while knowing it to be condemned by the author himself. What a noble and generous deed!'[33] Much more famous, of course, was the manipulation by Newton (and others working on his behalf) of many letters and related documents in the Society's possession – or elsewhere – in his war against Leibniz. Already in the 1690s we find David Gregory, Edmond Halley and John Wallis searching, among other places, the archives of the Society for evidence incriminating Leibniz for publication in Wallis's third volume of collected works. Much of the same material, augmented by documents provided by William Jones, was later used by Newton himself when composing the *Commercium epistolicum*, wherein it was also advertised that 'the original letters (or drafts) used in this way remained (as evidence) in the possession of the Royal Society'.[34] Nor should we forget that the second part of Sprat's *History* was based on the archival material supplied by Oldenburg[35] or that the subject matter of the *Philosophical Transactions* and, even more conspicuously, of the *Philosophical Collections* was derived from the Society's archive.

I hope that even this brief survey demonstrates that record-keeping was built into the very concept of institutionalization of the Royal Society. Though the concept of such a repository did not originate with the Society, its members were the first – and for a long time the only ones – to actively engage in such a novel enterprise. Michael Hunter rightly commented on the significance of the Royal Society's 'virtual obsession with record-keeping'.[36] Mindful of the fragility of human records, and fully cognisant of the necessity to create a place for safekeeping the anticipated vast body of information augured by the new form of co-operative scientific research, the Fellows of the Royal Society regarded their archive – at least initially – as an indispensable cornerstone of their enterprise. Furthermore, in publicising the creation and meticulous maintenance of such an archive, the members also sought to entice scientists and inventors alike to entrust their labours to the Society for safeguarding, thereby both preserving their priority

33 Shapin and Schaffer, *Leviathan and the Air-Pump* (n. 24), p. 348.

34 W.G. Hiscock (ed.), *David Gregory, Isaac Newton and their Circle* (Oxford, 1937), pp. 6–8; A. Rupert Hall, 'John Collins on Newton's Telescope', *Notes and Records of the Royal Society*, 49 (1995), 71–8, on p. 71.

35 Publicly Sprat claimed that it was he who had selected the material for his narrative 'out of their *Registers*, and *Journals*, which I have been permitted to peruse'. Sprat, *History* (n. 1), p. 4. See also above, pp. 179–80. For the known facts concerning the composition of the *History* see Paul B. Wood, 'Methodology and Apologetics: Thomas Sprat's *History of the Royal Society*', *British Journal for the History of Science*, 13 (1980), 1–26; Hunter, *Establishing the New Science* (n. 4), pp. 45–71.

36 Ibid., p. 4.

claim and preventing their decline into oblivion.[37] Equally important, however, a successful archive helped alleviate certain of the Society's concerns about its own permanence, for not only did the continued flow of information assure its dynamism, but by serving as the agency for the advancement of knowledge, the Society – like any munificent patron – enshrined its fame in perpetuity. If the Society failed ultimately to achieve these grandiose archival objectives, the problem lay not in a flawed concept, but in an unpreparedness for the enormity of the enterprise it had undertaken. So, too, the absence of comparable models to emulate and a tradition of truly committed officers to guide and organise contributed to the gradual abandonment of unrealistic earlier schemes, and their eventual replacement with the more limited – and manageable – goal of simply preserving a more or less intact record of the Society's transactions.

[37] See, for example, Oldenburg's 1667 letter soliciting information from Sluse: 'Public registers are kept by them; in these it is the duty of the two Secretaries, to which they are bound by solemn oath, to enter everything which the Fellows of this Society, or the philosophers of other nations, or other skilful and learned men perform, observe, and communicate as a record in honor of each individual contribution which wit and labor deserve.' Oldenburg, *Correspondence*, iii, 537.

12

Image versus Reality: the Archives of the French Académie des Sciences

Christiane Demeulenaere-Douyère
and David J. Sturdy*

The value to the historian of any archive depends in no small measure on the questions to which he or she seeks answers; but as the researcher works in an archive and becomes familiar with its contents, new questions and new lines of inquiry arise. The relationship between researcher and archive is dynamic: the potentialities of an archive expand in proportion to the range and sophistication of the issues investigated, and the historian correspondingly is stimulated to ask ever more questions as he immerses himself in the documentation. The subject of this essay – the archives of the French Académie des Sciences for the pre-Revolutionary period – presents features of great complexity which render the task of a historical assessment exceedingly demanding.

Given the diversity of the archives, the following discussion must be selective in its approach. It will remain within two principal boundaries. One is chronological. Although there will be reference to the entire period between 1666 and 1793 – the lifetime of the old Académie des Sciences – the essay will concentrate on the decades before the reform of 1699; that is, on the first phase of the history of the Academy. The second concerns the sources to be discussed. For reasons which will become apparent, the focus of attention will be two sets of papers in the archives of the Academy: the *procès-verbaux* and the laboratory records preserved by Claude Bourdelin.

Through these and related sources a variety of questions will be approached: to what extent do the archives indicate that the work of the Academy was conducted by individuals as against groups of academicians; what was the Academy's role in protecting intellectual property rights; who decided programmes of research (individual academicians, patrons, the Academy as a whole?); what material support did the Academy provide for its members; and as regards the archives of the Academy, how and why were they formed, by whom, and to what uses did

* The authors would like to thank Professor Alice Stroup for her comments on an earlier draft of this paper.

academicians put them? One other point should be stressed in these introductory remarks. Although much of this discussion will deal with the archives of the Académie des Sciences as presently constituted, it will range farther afield to include material which once was in the possession of the Academy but which, for various historical reasons, now is located elsewhere.

The Archives and their Dispersal

The researcher in any archive must be clear as to the scope of the material which it contains, and it is in this regard that some of the problems inherent in the archives of the Académie des Sciences become apparent, for they did not escape the consequences of the Revolution. In 1793 the National Convention suppressed all the royal academies, including the Académie des Sciences whose papers, books and *cabinet de curiosités* for the time being remained sequestered.[1] Fortunately they were not subjected to the wholescale destruction which befell the records of other *ancien régime* institutions in France. In 1795 the Institut national des sciences et des arts was founded and comprised three classes; the first class, 'Sciences physiques et mathématiques' (essentially the successor to the former Académie des Sciences), inherited the pre-1793 Archive and *cabinet de curiosités*. In 1805 the Institut moved from the Louvre across the Seine to its present location, known at that time as the Palais des Beaux-Arts and renamed 'Palais de l'Institut' in 1815. Because of shortage of space in the new premises, and in accordance with an internal *règlement* of the Institut (6 August 1796), the manuscripts and books of the former Académie des Sciences became the responsibility of the library of the Institut; and in 1807 a large proportion of the former *cabinet de curiosités* was distributed between the Musée du conservatoire national des arts et métiers and the Muséum national d'histoire naturelle (the former Jardin royal des plantes).

More depredations were visited on the archives of the Académie des Sciences by the Libri affair. In the 1840s Guglielmo Libri, member of the Académie des Sciences,[2] professor at the faculty of sciences in Paris and at the Collège de France, exploited his privileged position as secretary to a governmental commission charged with drawing up a general catalogue of manuscripts in Departmental libraries to steal thousands of documents which he then sold abroad clandestinely. He also stole from the Institut, and included in his haul were numerous packets of papers which once belonged to the Académie des Sciences. Along with the rest of

1 This passage is based on Christiane Demeulenaere-Douyère, 'Esquisse d'une histoire des Archives de l'Académie des Sciences', in Eric Brian and Christiane Demeulenaere-Douyère (eds.), *Histoire et mémoire de l'Académie des sciences: Guide de recherches* (Paris, 1996), pp. 45–54 and E.S. Saunders, 'The Archives of the Académie des Sciences', *French Historical Studies*, 10 (1978), 696–702; on the collections of the academy see the unpublished thesis by Camille Frémontier, 'Le cabinet de l'Académie des sciences (1666–1864)' (Paris IV–Sorbonne, 1994).

2 In 1816 the classes of the Institut resumed the nomenclature of 'Académies', hence the title 'Académie des sciences' was restored.

his loot they were recovered in the 1880s and were deposited in the Bibliothèque Nationale (the Académie des Sciences as yet did not have a fully operative *service des archives*); it is to the Bibliothèque Nationale that the historian still must go to consult them.[3]

Other Relevant Archives

Mention of the Bibliothèque Nationale raises the question of other depositories which contain documents directly relevant to the pre-1793 Academy. The Paris observatory falls into this category. Built in the early 1670s as an integral part of the Academy, the observatory was regarded by Colbert, founder and first protector of the Académie des Sciences, as the location to which that body might move its headquarters in preference to the confined premises at the Bibliothèque du roi in the rue Vivienne. This is not the occasion to retell the troubled history of the design of the observatory and its failure to become a Baconian 'Temple of Salomon' as Colbert hoped.[4] The point is that it developed as a centre of exclusively astronomical rather than general scientific research, and its archives correspondingly remain mainly technical in character. They contain copious details of observations conducted by academicians such as the Abbé Jean Picard, the Cassinis, the La Hires and Joseph-Nicolas Delisle.[5]

Much documentation directly related to the Académie des Sciences is to be found in the Archives Nationales.[6] The Academy was created by the state and throughout its history down to 1793 came under the aegis either of the royal household ('la Maison du Roi') or of various ministries. The Académie des

3 The Libri scandal (his full name: Guglielmo Brutus Icilius Timeleone Libri-Carrucci-Dalla-Sommaja [1803–69]) is outlined in the introduction to Léopold Delisle, *Catalogue des manuscrits des fonds Libri et Barrois* (Paris, 1888) and id., 'Vente de manuscrits du Comte d'Ashburnham', *Journal des savants* (June 1899), pp. 317–37. The most recent study is: P.A.M. Ruju and Marco Mostert, *The Life and Times of Guglielmo Libri (1802–1869)* (Hilversum, 1995). Papers of the Académie des sciences and its members stolen by Libri and deposited in the Bibliothèque Nationale after their recovery, are: nouvelles acquisitions latines 2340–2341 and nouvelles acquisitions françaises 5175 (papers of Roberval); n.a.fr.5133–5149 (analyses of experiments in the laboratory of the Academy, 1667–99); n.a.fr.5150–5153 (miscellaneous papers of the Academy); n.a.fr.5156, 6200 (papers of the Cassinis); n.a.fr.5162 (papers of Frénicle de Bessy); n.a.fr.5166 (papers of Lagrange); n.a.fr.6197 (astronomical observations by academicians, seventeenth and eighteenth centuries).
Other scientific papers recovered from Libri are: n.a.lat.1636–1638 (observations and correspondence of Gassendi); n.a.lat.1639–1642 (correspondence of Hevelius); n.a.lat.1643–1644 (papers of Viète); n.a.lat.2339 (papers of Fermat); n.a.fr.4507 (papers of Leibniz); n.a.fr.5160–5161 (papers and correspondence of Descartes); n.a.fr.5176 (papers of Mersenne and Pascal); n.a.fr.6202 (letters of Huet, Bishop of Avranches, including correspondence with Oldenburg, Justel, Huygens, Boulliau etc.); n.a.fr.6204–6206, lat.2338 (papers and correspondence of Mersenne).

4 See Antoine Picon, *Claude Perrault ou la curiosité d'un Classique* (Paris, [1988]), pp. 197–223, and C.J.E. Wolf, *Histoire de l'Observatoire de Paris* (Paris, 1902), pp. 1–10, 113–19, 194–5.

5 The references in the Archives de l'Observatoire are: Picard, D1, 14–16; the Cassinis, D1, 1–8, 13–21; D3, 1–30; D4; 1–29, D5, passim; the La Hires, D1, 18–20, 93, 94, 97; D2, 1–10; Delisle, C2, 14–15, 113–14.

6 See Guy Beaujouan, 'L'histoire des sciences aux Archives Nationales de Paris', *Archives internationales d'histoire des sciences*, 3 (1950), 874–81.

Sciences was linked organically to the crown, and the Archives Nationales are rich in documentation relevant to the protection and direction which the government afforded the Academy, including the observatory. The papers of the royal household contain much material relevant to the difficult question of the financing of the Academy. Colbert supported the Academy with subventions from the royal treasury: academicians thereby received occasional *gratifications*, bought research instruments, and equipped the laboratory of the Academy at the direct charge of the crown. The cost of building and equipping the Paris observatory likewise was met entirely by the royal household.

This policy of underwriting capital and current expenditure, conferring *gratifications* and meeting the research expenses of the Academy was continued, albeit at lower levels, by Louvois and Pontchartrain, Colbert's successors as protectors of the Académie des Sciences. However, it is not always easy to trace the contours of this expenditure for the later seventeenth century. Although the financial papers of the royal household were published by Jules Guiffrey at the end of the nineteenth century,[7] they do not tell the whole story. Moreover, in 1691 funding for the Academy was removed from the purview of the royal household and was excluded from Guiffrey's accounts for post-1691 years. Researchers interested in the later period therefore must investigate alternative sources in the Archives Nationales, many of which are scattered throughout the numerous series which comprise the archives, and are difficult to analyse even when they have been identified.

Financial data for the later seventeenth century has been published in recent years and additional research on Colbert's financial administration of the Academy is being carried out using sources at the Bibliothèque Nationale.[8] Indebted as we are to scholars conducting this research, we are still far from possessing a comprehensive understanding of the financial context within which the Académie des Sciences functioned during its first few decades. Likewise with respect to the eighteenth century, we still await a systematic study of the finances of the Academy. While the search for data must begin with the accounts of the royal treasury found in series G7 at the Archives Nationales, the archives of the Académie des Sciences also contain important material. First, there are the records of the Comité de Trésorerie, created in 1725 to oversee the reimbursements claimed by academicians as well as verify the accounts of the Academy's treasurer; only one register of the proceedings of the committee survives and it covers the period from 1761 to 1776. Secondly, Antoine Lavoisier's papers include financial information from his own term as treasurer (1791 to 1793), as well as financial records from the treasurerships of Buffon (1744–88) and Tillet (1788–91).

Still in the Archives Nationales, the records of various *ancien régime* ministries

7 Jules Guiffrey (ed.), *Comptes des Bâtiments du Roi sous le règne de Louis XIV* (5 vols., Paris, 1881–1901).

8 Alice Stroup, *Royal Funding of the Parisian Académie Royale des Sciences during the 1690s* (Transactions of the American Philosophical Society, 77, part 4, 1987); M.-J. Tits-Dieuaide, 'Le financement de l'Académie des sciences de 1666 à 1699', in Brian and Demeulenaere-Douyère (eds.), *Histoire et mémoire* (n. 1), pp. 351–3.

– especially those dealing with finance, agriculture, commerce, industry and public works – contain correspondence and other papers pertinent to the Académie des Sciences, while the archives of the marine, which are stored at the Archives Nationales, include much information on the contribution of academicians to the search for solutions to navigational and other maritime problems. Finally, it should be noted that in the Archives Nationales are miscellaneous items which appear to have been deposited there by accident rather than according to any preconceived design; they include, for example, the papers of individual academicians such as Jean Truchet ('le Père Sébastien') and Gilles Filleau des Billettes.[9]

Other archives to which the historian can turn are the Archives du ministère des affaires étrangères and the Archives historiques du ministère de la guerre, the latter being particularly rich on the relationship between the Academy and the army; and we should recall that among the early members of the Academy, Antoine de Niquet and Nicolas-François Blondel served as royal military engineers, and Louis Gayant, a surgeon in the army, was killed at Maastricht in 1673 while on active service. As regards the present-day location of the papers of individual academicians, it is often a matter of chance as to where they are to be found. Some are in the Académie des Sciences, but others are elsewhere; for example, the Muséum national d'histoire naturelle holds the papers of Tournefort, the Jussieu family, Isnard, Vaillant, Chirac and Rouelle.

Finally, a tangential claim to be included in a discussion of the eighteenth-century Académie des Sciences (although it lies outside the range of the present exercise) comes from the archives of the Société Royale des Sciences at Montpellier. Founded in 1706, the Société Royale legally was an extension of the Paris Academy. It was the only purely scientific body in the provinces (the others were multidisciplinary[10]) and was modelled on the parent body in Paris. The research of its members occasionally was published in the *Histoire et mémoires* of the Académie des Sciences, and the members of the two assemblies had mutual right of attendance at meetings.[11]

9 Archives Nationales [hereafter AN], M803, 849, 851.

10 On the spread of provincial academies and their character, see Daniel Roche, *Le siècle des lumières en province: académies et académiciens provinciaux, 1680–1789* (2 vols., Paris, 1978).

11 The archives of the Société royale des sciences are in the Archives Départementales de l'Hérault, D 116–230; see also, Junius Castelnau, *Mémoire historique et biographique sur l'ancienne Société Royale des Sciences de Montpellier* (Montpellier, 1858); Louis Dulieu, 'La Société Royale des Sciences de Montpellier (1706–1792)', *Actes du 100e Congrès national des sociétés savantes, Paris, 1965: Section d'histoire moderne et contemporaine* (2 vols., Paris, 1976–77), i, *Les sociétés savantes, leur histoire*, 43–48; and Eugène Thomas, 'Notice sur l'ancienne Société des sciences et l'ancienne Société des sciences et belles-lettres de Montpellier', *Annuaire administratif, historique, statistique et commercial de l'Hérault pour l'année 1859* (Montpellier, 1859), pp. 52–76.

The Content of the Archives of the Académie des Sciences

The present-day archives of the Académie des Sciences therefore constitute but part, important as it is, of the totality of sources relevant to a historical study of the Académie des Sciences. What types of material do they possess for the pre-1793 period? The recently-published guide to the archives[12] is the essential tool to be employed since it includes historical essays as well as comprehensive archival and bibliographical information.

With respect to records of the meetings of the Academy there are three sources to be considered: the *registres des procès-verbaux*, the *plumitifs* and the *dossiers de séance* (also called *pochettes de séance*). The *procès-verbaux* are the official minutes of the meetings and in their totality they cover the period from 1666 to the present day. Researchers have access to the *procès-verbaux* in the form of photocopies available on open shelves in the reading room of the Service des Archives; there is also available an index of the proper names which are found in the registers for the years 1666 to 1706. For the post-1795 period, the *procès-verbaux* are in print.[13] The *procès-verbaux* were written in French, although occasionally academicians or their correspondents presented papers in Latin. The *procès-verbaux* also changed in character over time. At first there were two sets of registers, one for 'les mathématiques' (comprising 6 volumes) and one for 'la physique' (comprising 4 volumes); thus for the period from 22 December 1666 to February 1684, there were ten volumes. No registers survive for the years 1670 to 1674.[14] In February 1684 the Academy began keeping only one set of registers for its meetings, a policy that has continued to the present day: thus for the period from 26 February 1684 to the end of 1698 there are seven volumes.

The *plumitifs* are minutes taken during the meetings of the Academy. They were then transcribed into the registers of the *procès-verbaux*, although there sometimes occurred delays of several years, a factor which explains lacunae, misspellings of proper names and other errors in the registers. The extant series of *plumitifs* covers the period from 1713 to the nineteenth century. None survive for the seventeenth century, but a perusal of the registers of the *procès-verbaux* suggests that they once existed. Finally there are the *dossiers de séance* formed by later archivists who

12 See Brian and Demeulenaere-Douyère, *Histoire et mémoire* (n. 1).

13 *Procès-Verbaux des séances tenues depuis la fondation de l'Institut [1795] jusqu'au mois d'août 1835* . . . (10 vols., Hendaye, 1910–22), with *Tables générales alphabétiques* (Paris, 1979) (Meetings from 6 nivôse an IV [27 December 1795] to 27 July 1835). From 1835: *Comptes rendus hebdomadaires des séances de l'Académie des sciences* [from 3 August 1835], with *Tables générales* of authors from 1835 to 1955 and subjects from 1835 to 1895.

14 These registers may have been destroyed, although the time and circumstances are unknown; possibly they were never written in the first place. However, that a *procès-verbal* of the meeting of 21 January 1673 once existed is indicated by a booklet, 'Machines nouvellement exécutées et en partie inventées par le Sieur Hubin, Emailleur ordinaire du Roy', Huygens, *Oeuvres complètes*, xix, 345, note 3. Prof. A. Stroup has informed the authors that some academicians claimed that papers read during 1674 were inserted into the register of 1675 (Archives de l'Académie des sciences, procès-verbaux [hereafter AdS, PV], vol. 7, fols. 43v (18 April 1676), 259 (18 January 1674); ibid., vol. 8, fols. 4v–7v (23 January 1675, 28 February and 28 March 1674).

attempted to collate original documents, letters, papers and reports presented to the Academy, in an effort to reconstruct the records of individual meetings. This series runs from 1667 to the present day.

These are the major categories of documentation dealing with meetings of the Academy, but there are others which should be noted. They include notebooks of the laboratory of the Academy dating from the seventeenth century, the papers submitted for prizes, sealed notes (*plis* or *dépôts cachetés*) from the end of the eighteenth century to the present, and miscellaneous documents not yet clearly identified relating to the Academy during the seventeenth and eighteenth centuries. There are also about sixty-five personal archives, or *fonds d'archives personnelles de scientifiques*, containing papers of individual academicians such as Roberval, Réaumur and Lavoisier. The Academy archives also possess manuscripts which have been bought or accepted as gifts; they include the *Journal d'Angleterre*, which is a translation into French of selected articles from the *Philosophical Transactions* of the Royal Society. There is also a personal file (a *dossier biographique*) on each member of the Academy; the files contain letters, manuscripts, miscellaneous papers, historical articles and other material pertinent to the individual academicians.

For all the historical vicissitudes which have affected their composition, the archives of the Academy nevertheless display remarkable lines of continuity. In spite of the consequences of the suspension of the Academy in 1793 and the subsequent losses resulting from the Libri affair, the records of the academy exhibit an exceptional degree of unity and continuity by comparison with other French administrative archives from before and after the Revolution. The Academy archives may not enjoy the same unbroken accretion as, say, those of the Royal Society, but this would be an unrealistic comparison: the Royal Society never confronted the socio-political crises to which the Académie des Sciences had to respond, and its archives were never dispersed as were those of its French counterpart. In the French context, wherein institutional and ministerial archives often are radically different before and after the Revolution, those of the Académie des Sciences reveal a striking unity of character and purpose.

Two key questions arise: what light do the archives shed on the history the early Académie des Sciences, and to what extent do they illustrate the uses to which academicians themselves put them? This latter question in particular holds out potential for our understanding of 'the scientific revolution'. If we accept that western science during the seventeenth century was undergoing a fundamental shift from the 'ancient' to the 'modern', then it may well be that part of the 'modernisation' process is to be found in a new use of scientific archives by scientists themselves. The first group of documents to be discussed is the *procès-verbaux*.

The *Registres des Procès-Verbaux*

To appreciate the significance of the *procès-verbaux* we should recall that, at its foundation by the crown through the initiative of Colbert, the Académie des Sciences was provided with no formal regulations or statutes, those regulations not being forthcoming until 1699. The exact circumstances of the creation of the Academy remain something of an enigma. Although it is clear that Colbert persuaded Louis XIV that it would be an act commensurate with royal honour to establish and provide funds for a society whose interests were scientific, it is still not known exactly when or how the decision was taken; certainly the records of the Académie des Sciences contain no indication of a momentous 'birth'.[15] On the other hand, the *procès-verbaux* imply that in some form a group of 'mathématiciens' (astronomers, geometers and physicists) was already functioning as early as June 1666;[16] over the next few months that core was enlarged by the addition of 'physiciens' (anatomists, botanists, chemists, physicians and apothecaries). From June to December 1666 this growing band of *savants* assembled in Colbert's Parisian *hôtel*, but by the end of the year was given rooms in the Bibliothèque du Roi.[17]

When the members of the Academy first formally gathered in December 1666, they belonged to a body with no name (for several years they referred to themselves as 'la Compagnie'), no organisation, no structure and no declared purpose. For several weeks they debated questions of procedure, the range of their scientific research, the principles underlying those activities, and a host of other related matters which may be traced through the early volumes of the *procès-verbaux*. However, a word of caution is necessary: the first two volumes of the *procès-verbaux* are somewhat confused and should be read concurrently in order to make sense of the years from 1666 to 1668. Those early debates are of the greatest importance to the historian, for they constitute an eloquent expression of the thoughts of the members as to their conduct as scientists who now also were academicians; to what extent, therefore, did this latter condition have implications for the former, and how did the academicians envisage their corporate activities? Their resolutions included the decision that the secretary would keep minutes of the bi-weekly meetings of the Academy, and that he would do so in French; it was these minutes which formed the basis of the *procès-verbaux*. During meetings the secretary would jot down notes on sheets of paper – the *plumitifs* referred to above – whose contents later were entered formally into the registers of the *procès-verbaux*.

15 René Taton, *Les origines de l'Académie royale des sciences* (Paris, 1966); Stroup, *Royal Funding* (n. 8); Alice Stroup, *A Company of Scientists. Botany, Patronage and Community at the Seventeenth-Century Parisian Royal Academy of Sciences* (Berkeley and Los Angeles, 1990); D.J. Sturdy, *Science and Social Status. The Members of the Académie des Sciences, 1666–1750* (Woodbridge, 1995); M.-J. Tits-Dieuaide, 'Une institution sans statuts: l'Académie Royale des sciences de 1666 à 1699', in Brian and Demeulenaere, *Histoire et mémoire* (n. 1), pp. 3–13.

16 AdS, PV, vol. 2, fols. 20 et seq.

17 Guy Meynell, 'The Académie des Sciences at the rue Vivienne, 1666–1699', *Archives internationales d'histoire des sciences*, 44 (1994), 22–37.

From the beginning, the academicians were emphatic that there must be preserved both for themselves and for posterity a formal and authoritative record of meetings. The *procès-verbaux* accordingly contain not only information on the scientific papers read at meetings and the debates which followed them, but also on the institutional and organisational decisions which the academicians reached. The *procès-verbaux* were the medium through which the academicians fulfilled the function which Colbert in effect assigned to them: to take upon themselves responsibility for the organisational and institutional principles governing the corporate activities of the Academy. The *procès-verbaux* embodied, as it were, the elements of a 'constitution' of the Academy; not in the sense that members laid down their own statutes, but in the sense that the *procès-verbaux* were the authoritative and cumulative record of the scientific, procedural, organisational and institutional programmes and measures which the academicians adopted over several decades.

What was the nature of the early procedural debates in the Academy as revealed by the *procès-verbaux*? The Academy had twenty-one members in December 1666, of whom over half were in their forties, fifties or sixties; another two – Richer and Huygens – were in their thirties. A perusal of the *procès-verbaux* for the late 1660s suggests that the early debates on the structure and scientific programmes of the Academy were influenced strongly by the more mature members, and especially by Huygens. He had been closely involved in the composition of the famous 'Project de la Compagnie des sciences et des arts' (a document composed probably in or shortly after 1663) which had offered Colbert a blueprint for a possible government-sponsored learned society,[18] and the scientific and organisational course taken by the Académie des Sciences during its early years seems to have corresponded closely to the ideals of Huygens. Even Claude Perrault, who also exerted a marked influence on the debates of 1667 and 1668, looks to have followed the line taken by Huygens.

In 1667 the members took several strategic decisions. One of the first, in January 1667, affirmed that the content of meetings would be kept secret.[19] Papers would be read and discussed, and experiments conducted or demonstrated, but nothing would be divulged to the public. The minutes of meetings, and therefore decisions affecting the evolution of the Academy, remained available only to the academicians themselves and, of course, their protector Colbert. To the decision on secrecy was added another of 16 March 1667: if a member made a new scientific discovery, then it must be scrutinised and confirmed by the Academy as a whole before being announced in the *Journal des sçavans*; and, although attributed to the academician in question, the discovery would be presented as an achievement of the entire Academy.[20] In both institutional and scientific spheres, the academicians placed strong emphasis on their corporate identity: the outer world would be informed of

[18] Sturdy, *Science and Social Status* (n. 16), pp. 70–2.
[19] AdS, PV, vol. 1, fol. 1.
[20] Ibid., vol. 2, fos. 159–60.

the collective decisions of the Academy, but not of the processes by which those decisions were reached.

This concern for corporate identity extended to the scientific programmes of the Academy. The *procès-verbaux* provide evidence that the scientific ethos created by the senior academicians predominantly was one of Baconianism.[21] Bacon's works had been available in French translation since at least 1619, but his ideas on the creation, organisation and dissemination of knowledge had also been publicised in the 1640s and 1650s through the literary *salons* and learned *cabinets* of Paris. Several of the founder-members of the Academy had been *habitués* of learned gatherings, and had acquired a Baconian attachment to experimental learning, programmes of collaborative research, and the careful preservation of scientific papers which would provide material for the compilation of 'Tables of Invention'. Members of the Academy sensed that Baconian strategies were especially appropriate to a scientific society. Baconianism proved compatible with the concept of 'probabilism' to which many academicians were attached (i.e. the notion that although science may not achieve 'absolute truth', it can aspire to 'probable truth' by minimising error through team-work); and Baconian strategies made possible the adoption of research projects on a scale which no individual could sustain. Hence there occurred the famous debates of 1667 which led Huygens to propose, and Perrault to develop, the plan for a natural history of plants.

Botany appeared to be well-suited to a Baconian plan. Teams of botanists would gather specimens throughout France, and perhaps even abroad; those specimens, and the scientific papers in which they were described, would be used by the Academy to publish a general *Histoire naturelle des plantes*. Professor Stroup's account of the history of this ambitious scheme is a cautionary tale.[22] At first the project involved mainly Perrault, Huygens and Duclos, but as other academicians became involved, notably Dodart and Nicolas Marchant, disputes arose over scientific method, and personal relations declined. Duclos in particular resented the presence of Dodart – thirty-five years his junior – and consistently criticised the quality of Dodart's research, questioning his scientific integrity and dismissing the papers which he wrote as unworthy of the Académie des Sciences.[23] Dodart maintained a dignified silence in the face of what must have been a most dispiriting hostility. As the years went by, immense quantities of paper pertinent to the *Histoire* were accumulated. Dodart edited an interim report in 1676 – *Mémoires pour servir à l'histoire des plantes* – whose preface, faithful to its Baconian inspiration, listed the academicians whose research had been used in its compilation. The process of accretion continued, the papers eventually devolving to Tournefort; but when he died in 1708 they were in a state of such confusion that

21 This is not to say that it was *exclusively* one of Baconianism. Cartesian philosophy looks to have influenced several academicians, although overt expressions of Cartesian sympathies were difficult because of the heterodox nature of Cartesianism; on the parallel influence of Baconianism as a 'natural' method for the Royal Society, see Michael Hunter, *Establishing the New Science: the Experience of the Early Royal Society* (Woodbridge, 1989), pp. 10–11.

22 Stroup, *A Company of Scientists* (n. 16).

23 Ibid., pp. 83–7.

nobody could fully sort them out, and the project was never completed. This unhappy history of the *Histoire naturelle des plantes*, which may be traced through the *procès-verbaux* and other related sources, is an instructive case-study of the difficulties inherent in collaborative science. Some of the problems were structural and methodological, but others related to questions of personality.

Such unflattering realities might be apparent to the sceptical eye of the modern historian, but to the academicians of the early 1700s looking back on the early *procès-verbaux*, it was the corporate identity of their predecessors which called for recognition. In 1733 the Academy published eleven retrospective volumes of its *Histoire et mémoires* covering the period from 1666 to 1699. Compiled from the *procès-verbaux* and from papers then in the possession of the Academy but subsequently lost, these volumes were used not only to present to the reading public the scientific work of the early academicians, but to explain the scientific strategy which they had developed. The first volume contained Fontenelle's famous history of the origins and early development of the Academy, but taken together the eleven volumes defend the proposition that one of the purposes of the programmes of experimentation of the 1666 Academy was to dispel individualism in favour of collaboration. The value of the papers of individual scientists allegedly was conditioned by the contribution which they made to the communal enterprise. Even though a present-day student of the *procès-verbaux* might conclude that the co-operative projects of the Academy often provided a convenient smokescreen within which academicians could pursue their individual research, the pretence of harmonious collaborative research was upheld by the eighteenth-century publications.

The preface to volume 3 of the 1733 *Histoire et mémoires*, which was Perrault's 'Mémoires pour servir à l'Histoire Naturelle des Animaux', put it thus:

> The most notable aspect of these *Mémoires* is their impeccable testimony to certain and recognised truth, for they are not the work of an individual . . . [Our *Mémoires]* contain only facts which have been verified by the whole Company, made up of people who have eyes for these sorts of things . . . and hands to seek them with greater dexterity and success; who see very well what is, and who could be made to see that which is not, only with difficulty; who study, not so much to find new things, as to examine closely those which were thought to have been discovered; . . . by so much does their love of certainty override all other considerations.[24]

The image of the seventeenth-century Academy created by the eighteenth-century

[24] 'Ce que nos Mémoires ont de plus considérable, est ce témoignage irreprochable d'une vérité certaine et reconnuë. Car ils ne sont point de travail d'un particulier . . . [Nos *Mémoires*] ne contiennent point de faits qui n'ayent été verifiés par toute une Compagnie, composée de gens qui ont des yeux pour voir ces sortes de choses . . . de même qu'ils ont des mains pour les chercher avec plus de dexterité et de succès; qui voyent bien ce qui est, et à qui difficilement on feroit voir ce qui n'est pas; qui ne s'étudient pas tant à trouver des choses nouvelles, qu'à bien examiner celles qu'on prétend avoir trouvées; . . . tant l'amour de la certitude prévaut dans leur esprit à toute autre chose.' (*Histoire de l'Académie des sciences* [Paris, 1733], iii, vii–viii).

body portrayed the early academicians as a noble company of scholars utterly devoted to scientific research, and soaring above the self-interest, jealousy and factionalism which too easily disturbed the corporate activities of other spheres of life. In practice, as the history of the *Histoire naturelle* revealed, members of the Académie des Sciences were no more immune to human failings and idiosyncrasies than anybody else, and they did not invariably conduct their scientific relations in a spirit of dispassionate inquiry. In short, the *procès-verbaux* may be used to sustain contrasting interpretations of the history of the Academy: the tendentious views of a Fontenelle and his colleagues, or the more objective analysis of a modern scholar.

As well as illustrating the evolution of scientific disciplines within the Academy, the *procès-verbaux* provide information on some of the less regular activities of the Academy. The bi-weekly meetings on Wednesdays and Saturdays occasionally had to be supplemented with special sessions: sometimes the opportunity arose to dissect the body of a hanged man or of an animal from the Versailles menagerie, and this had to be carried out over several consecutive days; or perhaps an eclipse occurred, requiring the attendance of the astronomers to conduct observations. Again, the Academy often delegated work to committees composed of a few academicians who met as and when it was convenient; they examined mechanical processes and machines submitted to Colbert or the Academy from all over France, and reported their conclusions to the Academy, Colbert or even to the king. The Academy also charged members with assessing recent books on scientific subjects, especially those published abroad. Meanwhile, in the 1670s several academicians found themselves teaching anatomy, chemistry and astronomy to the Dauphin; they presented accounts of these educational activities to the Academy, noting the satisfaction expressed by the royal family.

The *procès-verbaux* also testify to several amendments to the organisation and policies of the early Academy. The question of rules about publication recurred from time to time, and was discussed on at least eight occasions down to 1699.[25] Again, some time between 1670 and 1674, the order of meetings was reversed: thus, in the *procès-verbaux* for 1675, academicians changed the earlier pattern by treating 'la physique' on Wednesdays and 'les mathématiques' on Saturdays. Then, as we have already seen, it was decided to maintain a single register of *procès-verbaux* instead of having separate volumes for 'la physique' and 'les mathématiques'.[26] From 18 March 1684 the distinction hitherto preserved between life sciences and mathematical sciences was blurred; henceforth both 'la physique' and 'les mathématiques' were discussed on the same occasions. At the meeting of 18 March 1684, which normally would have been devoted to 'les mathématiques', reports were presented by Perrault and Du Verney on the natural history of animals

[25] AdS, PV, 15 January and 23 March 1667, 9 January 1676, 13 November 1686, 18 August 1688, 19 December 1691, 31 January 1692 and 26 January 1699.

[26] 'On mettra doresnavant dans un mesme registre et de suite ce qui a esté fait dans les assemblées de physique et de mathématiques parce qu'on a travaillé indivissemment sur des matières de physique et de mathématique' (AdS, PV, vol. 11, fol. 30v (26 February 1684)).

before the assembly turned to Cassini's paper on astronomical matters. The practice of combining life and mathematical sciences in the same meeting continued thereafter.[27]

Such changes represent not only the accumulated experience of the academicians, but also the wishes of their ministerial protectors. To this extent the *procès-verbaux* are a sensitive indicator of political influence upon the early Academy. It is evident, for example, that when Pontchartrain became protector in 1691 he resolved to impose what he considered to be badly-needed discipline. He insisted that members must work more diligently, and to this end required them to publish scientific papers more regularly than in the recent past.[28] Another measure which he introduced was to keep a record of attendance. Until 18 December 1694, the minutes of meetings simply opened with the formula: 'La Compagnie estant assemblée . . .'; from 18 December 1694 onwards, however, the wording became: 'L'assemblée étant composée de . . .', followed by the list of academicians present. When attendance at a particular meeting was small, the *procès-verbaux* would state the reason; for example, on 12 January 1695 cold weather was given as the explanation for a poor turn-out of members.[29] It was Pontchartrain who insisted that the registers of the *procès-verbaux* be kept more assiduously than in the past. Starting on 12 March 1695, he required that the folios of the registers be signed by the Abbé Bignon (president of the Academy) and by the Academy secretary as a guarantee that the *procès-verbaux* were written accurately and with suitable expedition.

The *procès-verbaux* are the most important single source in the archives of the Academy; they have been the basis not only of the research of nineteenth- and twentieth-century historians, but of that of eighteenth-century academicians who created an 'official' history and image of the early Academy. The *procès-verbaux*, although a confidential record available only to members of the Academy before the Revolution, nevertheless keep references to disputes to a minimum. For example, the controversy at the turn of the century between Méry and Du Verney over the passage of blood through the heart surprised the visiting English physician, Martin Lister, by its virulence; and he was shocked at the insults which the protagonists threw at each other. Yet the *procès-verbaux*, which include a reference to the lecture on this subject by Du Verney at the public meeting of the Academy in November 1699, make no mention of the acrimony with which the controversy was surrounded.[30] Even when he wrote the *procès-verbaux*, Du Hamel, the first secretary of the Academy, preferred discretion to frankness.

27 [Louvois required] 'que l'on travaillast au plustost à l'impression de l'Histoire des Animaux.' (ibid., vol. 11, fol. 52v (11 March 1684)).

28 Ibid., vol. 11, fol. 30v. However, in the 1670s the academicians had written more than the crown would allow them to publish. Stroup, *Royal Funding* (n. 9), pp. 50–1, argues that Pontchartrain economised on research and put extra resources into publishing work which academicians had already completed, in order to gain maximum publicity at minimum cost.

29 AdS, PV, vol. 11, fol. 30v. Stroup, *Royal Funding* (n. 9), ch. 4, table 4, discusses attendance as an index of declining morale.

30 Sturdy, *Science and Social Status* (n. 16), pp. 190–1, 294–5.

The Laboratory Papers of Claude Bourdelin

The second set of papers to merit special recognition is the records of the laboratory of the Academy between 1666 and 1699. Although the laboratory was designed by Duclos, who also supervised many of the programmes of experimentation, the records were written and kept by Claude Bourdelin.[31] The extant records are in several forms. First, there are the financial accounts which he kept.[32] They are a remarkable source on the equipping and running of a seventeenth-century laboratory, for they contain much more than lists of expenditure: they describe numerous experiments, many of them anatomical, and name the academicians for whom chemicals and equipment were bought. The entries in these volumes usually take the form of that for 28 June 1668: 'A lion was dissected, for which three "chopines" [1.5 litres] of brandy were needed, to wash the hands of the academicians and to pour into the internal organs, which were beginning to putrefy' (incidentally, the brandy cost 15 sous);[33] or again on 4 March 1670: 'A lioness was dissected, but stank strongly; a pint of brandy was used both to smell and drink, as well as to pour into the entrails' (this brandy cost 10 sous).[34]

The next set of laboratory books comprises eleven volumes written by Bourdelin and describing chemical experiments carried out between 1672 and 1699; there are no financial details in these volumes.[35] They are arranged chronologically, and the progress of the experiments is traced day by day. Thus, it is reported that, 'Tuesday 4 April 1673 there were distilled in a glass retort 2 pounds of wild narcissus without leaves'; there are notes on the progress of the experiment until it was completed on 7 April, when the results were recorded. A further note was added: 'Wednesday 26 April 1673 this plant was examined by Messieurs Perrault, Duclos, Dodart and Borelly. Seen and verified 26 April 1673'.[36] Most experiments in these volumes carry similar appendages to the effect that they have been reviewed and authenticated by other academicians; the last page of most volumes is signed by Jean-Baptiste du Hamel, secretary to the Academy, as a further guarantee of their reliability and authenticity. Du Hamel conferred, as it were, the imprimatur of the Academy.

31 On the laboratory see Joseph Schiller, 'Les laboratoires d'anatomie et de botanique à l'Académie des sciences au XVIIe siècle', *Revue d'histoire des sciences*, 17 (1964), 97–114.

32 They are now in the Bibliothèque Nationale, n.a.fr.5147–5149.

33 'L'on a dissequé un lion ou lon a employé trois chopines deau de vie, tant pour laver les mains de plusieurs que pour verser aux partyes internes qui comencoient à se gaster' (BN, n.a.fr.5147, fol. 7v).

34 'On a dissequé une lione qui sentoit tres mauvais, cest pourquoy on a donné une pinte deau de vie tant pout sentir que pour boire et pour jetter dans ses entrailles' (Ibid., fol. 27v).

35 AdS, pochettes de séance, cartons 1–2 (N.B. The Duveen-Lavoisier Collection, which forms part of the Rare Book and Manuscript Collections at the Carl A. Kroch Library, Cornell University, includes a copy in 6 volumes of 'Analises Chymiques de Monsieur Bourdelin Commencés le trente aoust 1672 et finies le deuxieme septembre 1699').

36 'Le Mardy 4e avril 1673 lon a fait distiler dans une cornue de verre 2ll. de simples fleurs de narcisses et sauvage sans aucun meslange de verdure . . . le mercerdy 26e avril cette plante a esté examinée a la présence des Messieurs Perrault, Duclos, Dodart, et Borelly. Vu et vérifié le 26 avril 1673.' (Ibid., carton 1, cahier 1, pp. 17–18).

The remaining sets of laboratory books comprise ten registers written probably by Dodart which arrange and analyse Bourdelin's aforesaid experiments alphabetically by subject (with cross-references to Bourdelin's records). Dodart's notebooks constitute a memorial to his attempts to correlate the results of Bourdelin's experiments as obtained by distillation, with the medical and nutritional value of plants.[37] Finally, there is a three-volume manuscript compiled in 1749 by Duhamel du Monceau, which also lists alphabetically Bourdelin's chemical experiments between 1672 and 1699; it replicates much material in Dodart's compilation.[38]

The implications of these laboratory books for our understanding of the working methods of the academicians are considerable; they have been explored by Claire Salomon-Bayet, and it is upon her analysis that this passage will chiefly draw.[39] First, we may consider the laboratory of the Academy as the focal point of its corporate scientific activities. The laboratory was, above all, a locus of experimentation; it symbolised the commitment of the Academy to experimental as against purely speculative learning, and to this extent implied that the Academy had adopted an anti-Aristotelian position in its approach to the study of nature and matter. The experiments undertaken in the laboratory were conducted under the aegis of the Academy, and the careful notes which were kept by Bourdelin (and possibly others) constituted a formal record over which the Academy could stand. In other words, the laboratory books legitimated the programmes of experimentation conducted on the premises of the Academy, and the experiments in return defined the Academy as a body committed to experimental learning.

The laboratory provided a context within which academicians, especially in the life sciences, could expand and refine seventeenth-century notions as to what was meant by 'experiment'. The term was used frequently in the seventeenth century (and formed part of the title of the Accademia del Cimento), but often with ambiguity: it could mean anything from the elementary exercises of a part-time alchemist, to the most sophisticated principles of scientific method as expounded in Cartesian or Pascalian thought. The laboratory of the Academy, it may be said, afforded members the opportunity to experiment with experimentation, especially as the Academy's commitment to secrecy protected members from potentially hostile scrutiny. Since the crown bore the costs of equipment and materials, it was possible to engage in sustained programmes of experimentation, to work in the presence of other academicians, to hear their comments (the laboratory served in effect as a seminar room as well as a place of experimentation), to devise new methods of attesting the validity of scientific hypotheses, and thereby to impose upon the term 'experiment' more rigorous meanings than in the past.

In short, the long-term epistemological implications of the experiments conducted in the laboratory were no less significant than their purely scientific content. The laboratory books, preserved by Bourdelin, examined and authenticated by

37 See Stroup, *A Company of Scientists* (n. 16), ch. 7.

38 AdS, 1J 18, i–iii.

39 Claire Salomon-Bayet, *L'Institution de la science et l'expérience du vivant: Méthode et expérience à l'Académie royale des sciences, 1666–1793* (Paris, 1978).

other academicians, and countersigned by the secretary of the Academy, therefore were more than routine accounts of experiments: they provided an accumulating 'fund' of experimentation on which members could draw, and were a statement of the scientific methodologies developed and authenticated by the Académie des Sciences.

Seventeenth-century scientists devoted much attention to the question of what constitutes 'proof' in science.[40] There was, for example, an increasing interest in the role of replicable experiments as a means of investigating particular problems and of demonstrating particular propositions. Some recent historical scholarship has contended that the question of scientific 'proof' contained a sociological dimension: the reliability of scientific evidence depended in part on the social status of those who presented it.[41] This particular thesis, developed in the context of a study of science in Britain, can be transferred into a French setting only with reservations. It is possible to argue that the first members of the Académie des Sciences were chosen in part with Colbert's eye on status: not so much that of the academicians themselves as the socio-political standing of their protectors. Gallois and Perrault were 'clients' of Colbert himself, while Carcavi had served the Duc de Liancourt, Pecquet had been attached to Fouquet (whose disgrace does not seem to have destroyed the prospects of his physician), Du Hamel had been advanced at certain stages of his career by Hardouin de Péréfixe, Archbishop of Paris and former tutor to Louis XIV, both Bourdelin and Marchant had served the king's uncle, Gaston d'Orléans, and Cureau de la Chambre had as his protector the Chancellor, Pierre Séguier.[42] The founder-membership of the Academy looks to have involved a balancing-act by Colbert who, in the interests of assuring the success of the Academy, selected sufficient 'clients' of powerful political figures to ensure the social respectability of the new body. As the decades passed, however, fewer such overtly 'political' appointments were made; membership of the Academy increasingly depended on scientific reputation or on the patronage of existing members of the Academy who brought younger scientific talent into the Academy. Membership of the Académie des Sciences guaranteed the probative authority of experiments conducted under its aegis; the social status of the individual academician consequently was irrelevant to the question of the value of his work.

40 See for example Steven Shapin and Simon Schaffer, *Leviathan and the Air Pump: Hobbes, Boyle, and the Experimental Life* (Princeton, 1985), and B.J. Shapiro, *Probability and Certainty in Seventeenth-Century England: A Study of the Relationship between Natural Science, Religion, History, Law and Literature* (Princeton, 1983).

41 Steven Shapin, *A Social History of Truth: Civility and Science in Seventeenth-Century England* (Chicago, 1994).

42 Sturdy, *Science and Social Status* (n. 16), pp. 139–40.

'Private' versus 'Public' Archives

To the historian, the significance of the *procès-verbaux* and laboratory books extends to the degree in which they shed light on the seventeenth-century concept of 'archives'. In the early modern period 'archives' certainly existed, but were deemed to be primarily depositories of legal documents: charters, letters patent, contracts and other papers defining rights, privileges and prerogatives. Correspondence or other material which had no direct bearing on the legal rights and status of an institution either was not preserved, or remained the property of the individuals who had written or received the letters or papers. It was not until the eighteenth century that this concept was modified; it was then, for example, that French ministerial archives began to preserve correspondence on a regular basis, augmenting their 'legal' function by the addition of a 'historical' dimension as material recording day-to-day activities was preserved.[43] The foregoing discussion of the *procès-verbaux* and laboratory books has implied that, from its foundation, the Academy inadvertently anticipated eighteenth-century developments. To the extent that they recorded procedural decisions the *procès-verbaux* in particular may be said to have served a juridical purpose, but through the accounts which they provide of the content of the meetings of the Academy they constitute the basis of an understanding of the historical evolution of the assembly.

The laboratory books raise the question of the ownership of scientific papers. Members of the Academy generally displayed a patrimonial attitude towards their scientific papers. This was borne out by the fact that when they bequeathed papers to the Academy they did so in the context of the general disposition of their personal estates, and the Academy accepted gifts in that spirit. This was affirmed as early as 1675 on the death of Roberval, who bequeathed to the Academy his mathematical papers and scientific instruments;[44] to the bequest he attached certain conditions which, again, the Academy recognised as binding.[45] The examples of other academicians who left papers, scientific instruments, or medical and other collections to the Academy,[46] confirm the principles observed in 1675. Academicians welcomed such gifts and carefully preserved them on the ground that they might contain unpublished information which could assist with their own research. The *procès-verbaux* for the meeting of 13 March 1683, for example, contain an inventory of Picard's papers which were given to La Hire.

43 See Amédie Outrey, 'La notion d'archives en France à la fin du XVIIIe siècle', *Revue historique de droit Français et étranger*, 2 (1953), 277–86; for a more general discussion of archives and their evolution in the early modern period, see Jean Favier, *Les archives* (Paris, 1959), chs. 1 and 2.

44 AN, Minutier Central, C–325, testament, 10 Oct. 1675.

45 The condition was that before any of the papers were published, they were to be corrected by Blondel, Buot and Picard. The Académie took possession of the papers in 1676 (AN, Minutier Central, XXIII–327, délivrance de legs, 23. Sept 1676); they are listed in the *procès-verbaux* of the Académie (see Appendix below).

46 Academicians who made bequests include, for example, Joseph-Guichard du Verney (1648–1730) and Louis Carré (1663–1711); on these two academicians and the details of their bequests see Sturdy, *Science and Social Status* (n. 16), pp. 192, 243.

However, this and other such instances of bequests should not be interpreted as evidence of a conscious policy to establish a *service des archives* operating according to modern principles of classification, organisation and preservation (this did not come until the nineteenth century), for no obligation was placed on academicians to leave their scientific papers to the Académie des Sciences. Rather do they constitute a practice of accumulating papers on an *ad hoc* basis according to the testamentary purposes of individual academicians. Returning to Bourdelin's laboratory books, circumstantial evidence suggests that the registers which are now in the archives of the Academy were considered to be the property of the Academy from the very beginning – in other words, Bourdelin acted on behalf of the Academy as he composed them;[47] on the other hand, the financial accounts which are in the Bibliothèque Nationale may well have been his own property, although a definitive pronouncement on this matter is impossible.

The habit of confusing private and 'academy' papers continued down to the Revolution. Among the provisions of the 'règlement' of 1699 which reformed the Academy were instructions concerning documentary and material records. The 'règlement' made references to the 'registres, titres et papiers' of the Academy, and ascribed to the secretary responsibility for maintaining them; in addition, at the end of every calendar year the president of the Academy was to make an inventory of the papers which he was entrusting to the secretary. These inventories are missing and it is even possible that they were never made. Nevertheless, the secretary had the duty of keeping the register of the *procès-verbaux* and other written records. The instruments, books, furniture and other collections which came into the possession of the Academy were made the responsibility of the treasurer.

Such arrangements seem straightforward enough, but their implementation was complicated by the absence of a clear distinction between 'academy' and private papers. Lavoisier, who was director of the Académie des Sciences in 1785, then its treasurer, kept among his own papers many 'mémoires' presented to the Academy, various submissions from candidates for the saltpetre prize, treasurer's records, and other material deriving from his activities within the Academy. It is only because the modern archives of the Académie des Sciences possess the Lavoisier papers that researchers have direct access to this immensely important source. Condorcet, who was the Academy's secretary from 1776 down to the dissolution of 1793, also failed to distinguish between his own papers and those of the Academy. The first act of the First Class of the Institut National des Sciences et des Arts in 1796 was to claim back the papers of the Académie des Sciences which Condorcet's widow had presented to the nation. An outstanding example of the primacy of private over institutional claims is the papers of the Abbé Jean-Paul Bignon. Introduced into the Academy in 1691 by his uncle, Louis Phélypeaux, Comte de Pontchartrain, he was alternately president or vice-president of that assembly until the 1730s. The changes introduced into the Academy during

47 An inventory of the papers of the Académie drawn up during the Revolution, An. II, refers to the various volumes just discussed (AdS, DG38), as does an inventory of 1807 (ibid.).

the 1690s, and above all the great reform of 1699, were largely his doing. He was deeply involved in the affairs of the Academy, and maintained an abundant correspondence with scientists and other scholars all over France, Europe, North America and China.[48] Yet the archives of the Académie des Sciences contain only the scantiest details on him; his vast correspondence and other papers remained his private possession, and are preserved as such in the Bibliothèque Nationale.[49]

The 'Règlement' of 1699

On 20 January 1699, the king presented the Academy with its first formal regulations; they were read out to the meeting of 4 February 1699. According to Fontenelle's minutes, 'Mr the abbé Bignon brought a règlement that the king had the goodness to establish for the Company. I read it at the meeting and here is the copy.'[50] Afterwards, a letter from the king nominating Bignon as president of the Academy was 'received by the entire group with an extreme joy'.[51] The regulations affirmed that the Academy was an official institution, with the title 'Académie royale des sciences'. The Academy depended directly on royal authority, usually represented by the minister of the royal household. The king assumed the role of protector of the Academy and guaranteed the funding of the experiments as well as pensions and attendance tokens (*jetons*) of its members. The regulations confirmed in formal terms some of the practices that had evolved since 1666, but they also recognised that in other respects there had been failures. One of the main ideals to be diminished in importance in 1699 was that of collaborative research; it was no longer regarded as a defining principle of the Academy's activities. The Academy now was recognised as a body of *savants* whose individual research would expand the sum of scientific knowledge. This was articulated in article 20 of the règlement:

> Experience having demonstrated too many inconveniences in the projects which the entire Academy undertook in common, each of the academicians will choose some particular object of his study, and in his account of these at meetings, he will attempt to enlighten the entire Academy, while at the same time profiting from their remarks.[52]

The meeting of 4 February was devoted entirely to the question of reform.

48 On his China correspondence see Claudia von Collani, *Eine Wissenschaftliche Akademie für China: Briefe des Chinamissionars Joachim Bouvet S.J. an Gottfried Wilhelm Leibniz und Jean-Paul Bignon über die Erforschung der Chinesischen Kultur, Sprache und Geschichte* (*Studia Leibnitiana*, 18 (1989)).

49 Bignon's papers are in BN, MS français 22225–22236.

50 'Mr l'Abbé Bignon a apporté un règlement que le Roy a eu la bonté d'arrester pour la Compagnie. Je [Fontenelle] l'ai lû à l'assemblée et en voicy la copie.' (AdS, PV, vol. 18, p. 107).

51 'ce qui a été receu de toute l'assemblée avec une extrême joye' (AdS, PV, vol. 18, p. 114).

52 'L'expérience ayant fait connoître trop d'inconvénients dans les ouvrages auxquels toute l'Académie pourroit travailler en commun, chacun des académiciens choisira plutôt quelque objet particulier de ses études, et par le compte qu'il en rendra dans les assemblées, il tâchera d'enrichir de ses lumières tous ceux qui composent l'Académie, et de profiter de leurs remarques' (Règlement of 1699, art. 20).

According to the new rules, the king would henceforth have the right to name the members of the Academy after they were nominated by the company itself. The academicians who were members before 1699 had their status confirmed. Stringent rules of residency and attendance were adopted, an exception being made for Denis Dodart, whose responsibilities as physician to the dowager Princess de Conti prevented him from residing in Paris (as required by article 4 of the règlement). Special circumstances were pleaded in his favour because he was a member of such long standing, with the reservation that the exception would not create any precedent.[53]

Subsequent meetings were devoted to nominating the members, settling procedural matters, and planning work. Such details were necessary because under the new rules the Academy grew in size from roughly thirty-four[54] to seventy members. Several categories of membership were formalised, ranging from honorary members to pensioners, students, associates and veterans. Organisational matters took up the whole of the meetings of 7, 11, 18, 21 and 27 February 1699; on 4 March 1699 the corresponding members were nominated. The ten *honoraires* were usually high functionaries or courtiers from whose ranks the king chose the president and vice-president of the Academy. The twenty *pensionnaires* included three geometers, three astronomers, three mechanicians, three anatomists, three chemists, three botanists, a permanent secretary and a treasurer; each received a pension from the king. Of the twenty *associés* twelve had to be *régnicoles* (i.e. subjects of the King of France) while the remaining eight could be foreigners. In addition, there were twenty *élèves*, each attached to a pensioner; after 1716, the student members were renamed *adjoints*. Except for the *associés*, who were not expected to attend the meetings of the Academy on a regular basis, academicians were required to reside in Paris. If they did not, they were liable to be replaced and moved to the category of *vétérans*. During these same weeks, the academicians determined such matters as the seating for pensioners at meetings (a hierarchy of seats was observed), the Academy's coat of arms ('a sun surrounded by three fleurs de lys arranged in the same manner as on the French écu'), and the research plans of individual academicians for the coming year, in accordance with the regulations.

Conclusion

The Académie des Sciences possesses one of the great scientific archives of the world. They serve as a reminder that, for all its internal dynamic, the practice of science is rooted in historical context. The archives of the Academy are an essential source not only for a study of the Academy itself, but for the whole movement of European science in the late seventeenth and eighteenth centuries. To the historian they are a depository whose potential is inexhaustible; as the enterprise of the history of science continues, the archives of the Académie des Sciences will

53 'Qu'à raison de son extrême ancienneté dans l'Académie et sans qu'un pareil exemple puisse dans la suite être jamais tiré à conséquence' (AdS, PV, vol. 18, p. 115).

54 Stroup, *Royal Funding* (n. 9), pp. 11–14.

remain one of the chief depositories to which scholars turn. They were exploited at the beginning of the eighteenth century by Fontenelle when he established the *Histoire et mémoires* and were central to his history of the Academy. Indeed, we may say that the archives of the Académie have fulfilled three main functions: they have safeguarded an important if not comprehensive record of the scientific activities of academicians from 1666 to the present day; they have preserved information on the institutional history of the Academy; and they have been the foundation of the written history of the Academy. The seventeenth-century Academy felt that the preservation of records was an integral part of its activities, not simply in order to provide later generations with useful scientific papers, but to establish the Academy as a living organism; an archive, in short, was necessary to the creation of a historical 'memory' of the Academy. The scientific papers and other materials preserved by the archives imply that academicians possessed a sense of obligation towards their successors who would enter into the heritage of academy life. That the archives continue to stimulate research is a testimony to the secretaries and others who compiled them; and there is every sign that we may anticipate new dimensions to the study of this great depository in the future.

Appendix
Bequest of Roberval to the Académie des Sciences

Archives de l'Académie des Sciences: procès-verbaux, Registre de Mathématiques, 5 Jan. 1676 – 2 Sept. 1679 [fol. 64v, 5 Dec. 1676]

Monsr. Picard ayant fait voir à la Compagnie l'Inventaire des papiers du feu Monsr. de Roberval, dont ensuit la Coppie, la Compagnie a arresté que ces papiers demeureront entre les mains du Sieur Du Hamel Secretaire de l'Académie, pour en prendre ceux dont ils auront besoin.

[fol. 65] Inventaire des manuscrits qui ont esté trouvez dans le Cabinet de feu Mr de Roberval.

Premier Paquet

A. Livre couvert de carton, contenant quelques calculs d'Eclipses; l'Explication de la Roulette, Considerations sur la Parabole, et quelques petits traittez des nombres.
B. Livre couvert de parchemin, contenant 17 Chapitres des triangles en nombres.
C. Livre couvert de carton, une suitte d'Apollonius, ou il est traitté des tactions de la Sphære, plus une Introd. aux lieux, plans et solides, plus des lieux à la superficie.
D. Livre couvert de carton. La division des figures planes.
E. Livre couvert de carton. de recognitione æquationum.
F. Livre couvert de carton. Leçons touchant l'Analyse Geom.
G. Livre couvert de carton. Divers fragments de Viete, avec 2 cayers detaches.

[fol. 65v]

H. Cayers des triangles spheriques d'Anderson.
I. Livre couvert de carton. Lieux plans, plus une methode de Max. et min.
K. Livre couvert de parchemin. Traittez d'Arithmetique, et de Geom. pract.
L. Livre couvert de carton. De la resolution des æquations planes et cubiques, de la main de Monsr. de Roberval.
M. Livre couvert de parchemin. De l'Analyse, contenant divers problemes tant plans que solides.
N. Livre couvert de parchemin, contenant divers traittez de Mr Frenicle.
O. Livre couvert de parchemin. Autres traittez de Monsr. Frenicle.
P. Livre couvert de carton. Geometrie des Indivisib. de Mr de Roberval.
Q. Livre en parchemin. Une Analyse Geometrique.
R. Livre en parchemin, deux traittez sur l'Analemme, plus un traitté de la coupe des pierres.
S. Livre en parchemin. Un traitté du Cone et du Cylindre, et de la sphere inscrite, ou circonscrite.
T. Livre de carton, traitté du Cone, et de la Sphere.

[fol. 66] Second Paquet

A. Livre de carton, traitté de la sphere.
B. Livre de papier bleu. Des quantites irrationelles.
C. Livres de carton. Divers prob. de Geom.
D. Deux copies d'une lettre latine de Mr de Roberval.
E. 36 cayers contenants des leçons de Mathem.
F. 4 cayers de Geom. pract.
G. 2 cayers de Mechanique, et de la conduitte des eaux, plus 8 cayers du mesme traitté.
H. Cayer de Chronol.
K. Cayer contenant 32 pages des mouvements composez, plus un autre traitté des mesmes mouvements pour les touchantes.
L. Cayer 12 pages de speculo vitoris par Anderson.
M. 8 feuilles, ou fragments de Dioptrique.
N. Feuilles contenant 17 prob. de Dioptrique.
O. Feuilles Elementa musicæ leçons.

[fol. 66v]

P. Calendrier.
Q. Du centre de percussion.
R. Traittée de Mechanique.
S. Un probleme de Mr Renaud de Marseille.
Feüille touchant le probleme de Cassius, resolu par Mr Descartes.
19 Cayers de Geom. pour les fortif.
Traitté de Musique.
Traitté de la Sphere imparfait [sic].

Troisiesme Paquet

L'Original de l'Aristarque en feüilles volantes.
Apoll. de locis planis 3 cayers.
Radicum secundi, et ulterioris ordin. usus un cayer
Ad locos plan. Isagoge 2 feüilles, plus 5 autres feüilles de mesme.
De doctrinâ tangentium 1 cayer.
Response à l'escrit des amis de Monsr. Fermat. 1 cayer.
Traitté d'optique en feuilles volants enfumé
3 Cayers de Dioptrique
Mechanique en feuillles, enfumé

[fol. 67]

5 cayers de Catoptrique
Du calcul des Eclipses pour la Chronologie 1 Cayer
Discours du vuide
De Centro virtutis potentiarum 3 feuilles
1 Cayer de Perspective
7 feuilles des Elements de Geom. de M. de Roberval
Fragments de Catoptrique
8 petits cayers d'Analyse liez ensemble

De Conis Isoperimetris etc. 3 cayers
De Superficiebus Cylind. in sphæra 3 Cayers
Solution d'un probleme tiré d'un livre holandois intitulé Wiss-Konstenaer
Del volare che non sia imposs.
1 Cayer de fortif.
6 Cayers de perspective
Recueil du calcul qui sert à la Geometrie six Feuilles
Fragments, ou feuilles cottées depuis α et jusques à X

Quatriesme Paquet
Deux lettres de Mr Hugens cottées à B
[fol. 67v]
29 Lettres de Mr des Noyers de Pologne.
7 Lettres de Mr de Beaune.
4 Lettres de Mr Hevelius.
10 Lettres apostillées de la main du Pere Mersenne.
4 Lettres de Mr le Tanneur.
2 Lettres de Mr Frenicle.
18 Lettres de Torricelle.
75 Lettres de Mr Descartes.
39 Lettres de Mr Fermat (les lettres de Mr Fermat données à Mr de Carcavi pour en tirer copie)
Liasse de lettres cottées depuis y, excepté une, jusques à C.

Cinquiesme Paquet
Les Elements de Geometrie de Mr. de Roberval,
Plus un traitté de la sphere, et du cylindre

Index